International
Association
of Fire Chiefs

By Dr. Ben A. Hirst,
Performance Training Systems

EXAM PREP

Fire Fighter I & II

SECOND EDITION

JONES AND BARTLETT PUBLISHERS
Sudbury, Massachusetts
BOSTON   TORONTO   LONDON   SINGAPORE

Jones and Bartlett Publishers
World Headquarters
40 Tall Pine Drive
Sudbury, MA 01776
978-443-5000
www.jbpub.com

Jones and Bartlett Publishers Canada
6339 Ormindale Way
Mississauga, Ontario L5V 1J2
Canada

Jones and Bartlett Publishers International
Barb House, Barb Mews
London W6 7PA
United Kingdom

International Association of Fire Chiefs
4025 Fair Ridge Drive
Fairfax, VA 22033
www.IAFC.org

Performance Training Systems, Inc.
250 South Central Boulevard
Suite 108
Jupiter, FL 33458
www.FireTestBanks.com

Jones and Bartlett's books and products are available through most bookstores and online booksellers. To contact Jones and Bartlett Publishers directly, call 800-832-0034, fax 978-443-8000, or visit our website www.jbpub.com.

Substantial discounts on bulk quantities of Jones and Bartlett's publications are available to corporations, professional associations, and other qualified organizations. For details and specific discount information, contact the special sales department at Jones and Bartlett via the above contact information or send an email to specialsales@jbpub.com.

Editorial Credits
Author: Dr. Ben A. Hirst

Production Credits
Chief Executive Officer: Clayton E. Jones
Chief Operating Officer: Donald W. Jones, Jr.
President, Higher Education and Professional Publishing:
 Robert W. Holland, Jr.
V.P., Sales and Marketing: William J. Kane
V.P., Production and Design: Anne Spencer
V.P., Manufacturing and Inventory Control: Therese Connell
Publisher, Public Safety Group: Kimberly Brophy
Senior Acquisitions Editor—Fire: William Larkin
Managing Editor: Carol E. Guerrero
Editorial Assistant: Laura Burns

Production Manager: Jenny L. Corriveau
Production Assistant: Tina Chen
Marketing Manager: Brian Rooney
Interior Design: Anne Spencer
Composition: Northeast Compositors, Inc.
Cover Design: Kristin E. Parker
Photo Research Manager and Photographer: Kimberly Potvin
Printing and Binding: Courier Stoughton
Cover Printing: Courier Stoughton
Cover Image: © Almir1968/Dreamstime.com
Unless otherwise indicated, photographs are under copyright
 of Jones and Bartlett Publishers, LLC.

The procedures in this text are based on the most current recommendations of responsible sources. The publisher and Performance Training Systems, Inc. make no guarantees as to, and assume no responsibility for the correctness, sufficiency, or completeness of such information or recommendations. Other or additional safety measures may be required under particular circumstances. This text is intended solely as a guide to the appropriate procedures to be employed when responding to an emergency. It is not intended as a statement of the procedures required in any particular situation, because circumstances can vary widely from one emergency to another. Nor is it intended that this textbook shall in any way advise firefighting personnel concerning legal authority to perform the activities or procedures discussed. Such local determinations should be made only with the aid of legal counsel.

6048
Printed in the United States of America
14 13 12 11 10   10 9 8 7 6 5 4 3 2

CONTENTS

ACKNOWLEDGMENTS

More than 14 fire department fire fighters have contributed to the development, validation, revision, and updating of the examination items included in this *Exam Prep* book. Their efforts have spanned more than 3 years and are valued because of the credibility they provided. A special thanks goes to the recent Technical Review Committees for Fire Fighter I and II for validating and updating the examination items to the National Fire Protection Associations (NFPA®) 1001, *Standard for Fire Fighter Professional Qualifications*, 2008 Edition Standard and latest technical publications: Chief Dave Brooks, City of DePere Fire Department, Wisconsin; Larry Swartz, Jr., Fire Service Program Director, Manatee Technical Institute, Bradenton, Florida; Ray Goff, Instructor, Evansville Fire Department, Evansville, Indiana; Deputy Chief Michael Jepeal, Simsbury Fire Department, Simsbury, Connecticut; Michael Jones, Fire Coordinator, Emergency Training Center, Crowley Texas and Fire Lieutenant, Burlesson Fire Department, Burlesson, Texas; Lieutenant Brian Hurst, Manchester Fire Department, Manchester Fire Rescue/EMS, Manchester, Connecticut; David W. Estes, Fire Fighter Engineer, Pike Township Fire Department, Indianapolis, Indiana; Jason Loyd, Instructor, TEEX-Texas A&M Systems, College Station; Texas; Chief Vicki Sheppard, Palm Beach County Fire Rescue, West Palm Beach, Florida; and Alan Joos, Assistant Director, LSU Fire & Emergency Training Institute, Baton Rouge, Louisiana. These individuals worked diligently to make considerable improvements in the examination items.

I want to thank my wife Ann, family, and friends who encouraged me to continue pressing forward with the work. Without their understanding and support, I would not have been able to meet the scheduled delivery.

Last, but not least, I express my sincere thanks to my able staff: Ellen Korn, Technical Editor; Beth Rimes, Administrative Assistant and Sales Support; Walter Hirst, Lt. Special Operations Training Officer and PTS Director of Operations; Chris Gass, Paramedic/Fire Fighter and PTS Regional Sales Manager; and Cathy Hemp, PTS Regional Sales Manager. While I was away, in complete solitude, they kept the business and the office operating at peak performance.

PREFACE

The Fire and Emergency Medical Service is facing one of the most challenging periods in its history. Local, state, provincial, national, and international government organizations are under pressure to deliver ever-increasing services. The events of September 11, 2001, continued activities and threats by terrorist organizations worldwide, recent natural disasters, and the need to maximize available funds are some of the reasons most Fire and Emergency Medical Service organizations are examining and reinventing their roles.

The challenge of reinventing the Fire and Emergency Medical Service to provide the first response efforts includes increasing professional requirements. Organizations such as the National Fire Protection Association (NFPA), National Professional Qualifications Board (Pro Board), International Fire Service Accreditation Congress (IFSAC), the International Association of Fire Chiefs (IAFC), the International Association of Fire Fighters (IAFF), and the Federal Aviation Administration (FAA) are having a dramatic influence on raising the professional qualifications of the first line of defense for emergency response.

Qualification standards have already been improved. Indeed, accreditation of training and certification are at the highest levels ever in the history of the Fire and Emergency Medical Service. These improvements are reflected in a better-prepared first responder, but also have a noticeable effect on those individuals who serve. Fire fighters are being required to expand their roles, acquire new knowledge, develop new and higher level technical skills, as well as participate in requalification and in-service training programs on a regular basis.

The aftermath of September 11, along with several major natural disasters, has profoundly affected the Fire and Emergency Medical Service. Lessons learned, new technology, and the national focus on terrorism and weapons of mass destruction are placing much greater demands on first responders to keep abreast of change in their specialty operations and improve their technical competence in new technology that was not even available just a few years ago.

Fire fighters cannot afford to be complacent and continue to perform in the same way. Obvious dangers faced by first responders under current heightened security conditions require many adjustments in what is being taught to fire fighters as they prepare to operate in an emergency environment. To meet this need, processes and modes of operation must be carefully examined and must be continuously monitored, changed, and updated.

The United States' national leaders constantly point to the first responders as our "first line of defense" against acts of terror and defense of life and property from extremely dangerous weapons that have never been used extensively in U.S. history. The Fire and Emergency Medical Service is steeped in tradition. Nevertheless, it is critical that we question our traditions and our thinking to bring our knowledge, skills, and abilities in line with the demands of today's real world.

Many things have been learned from the September 11 attack on America. Some of these lessons learned were the result of our reluctance to change processes and procedures (i.e., our traditions). As great as our traditions are, members of the Fire and Emergency Medical Service industry must not stop reflecting on the paramount reasons for our existence—namely, to protect property, save lives, and perform our tasks with personal safety as the number one concern. These are very important reasons to exist, to continuously improve, and to move from a good Fire and Emergency Medical Service industry to a great Fire and Emergency Medical Service industry.

A few words about knowing vital strategic and tactical information are in order. Many organizations focus a great deal of their training time and effort on the performance side of firefighting. That is essential and is the bottom line for developing skilled fire fighters. The dark side of this approach to training is that it fails to emphasize key knowledge requirements. Often, it is not what we did or didn't do as a first responder, but what we could have done if we had a strong base of knowledge that helps to analyze and detect a need for action outside our routine tasks. It takes a knowledgeable person to recognize the need to implement Plan B when Plan A isn't effective or efficient. The Fire and Emergency Medical Service as a whole, and each first responder in particular, must focus equally on the knowledge portion of emergency responses to help improve the performance side of our tasks. We in the Fire and Emergency Medical Service have a unique opportunity to provide valuable solutions to attacks on the United States, handling of chaotic situations such as weapons of mass destruction, serious acts of nature, and terrorist events whose tactics are not even known today.

Pre-planning for an emergency target hazard is important; knowing what can go wrong is essential. Knowledge is the potential power we must acquire to make important adjustments during the emergency incident, to continuously size up the situation, and to alter our plan of action as needed and required. Our fire officers, fire fighters, and medical personnel must develop a solid knowledge base so that better judgments, sizeups, and fireground actions become possible. Research in education and training over the years has concluded that lack of knowledge is one of the key reasons why tasks are poorly performed or performed in a manner that did not achieve the expected results. Lack of knowledge often results in poor or non-performance of critical emergency responses.

Fire fighters generally don't like to take examinations. For that matter, few people really like them. The primary purposes of the *Exam Prep* series are to help Fire and Emergency Medical Service personnel develop an improved level of knowledge, eliminate examination-taking fear, build self-confidence, and develop good study and information mastery skills.

In the past 22 years, Performance Training Systems, Inc. (PTS) has emerged as the leading provider of valid testing materials for certification, promotion, and training for Fire and Emergency Medical Service personnel. More than 40 examination-item banks, containing more than 22,000 questions, provide the basis for the validated examinations. All products are based on the NFPA Professional Qualifications Standards, FAA regulations, and the DOT National Curriculum for Medical First Responders.

Over the past 15 years, PTS has conducted research supporting the development of the Systematic Approach to Examination Preparation® (SAEP). SAEP has resulted in consistent improvement in scores for persons taking certification, promotion, and training completion examinations. This *Exam Prep* manual is designed to assist fire fighters in improving their knowledge, skills, and abilities while seeking training program completion, certification, and promotion. Using the features of SAEP, coupled with helpful examination-taking tips and hints, will help ensure improved performance from a more knowledgeable and skilled fire fighter.

All examination questions used in SAEP were written by Fire and Emergency Medical Service personnel. Technical content has been validated through the use of current technical textbooks and other technical reference materials; and job content has been validated by the use of technical review committee's representative of the Fire and Emergency Medical Service ranks, training academies, and certification organizations. The examination questions in this *Exam Prep* manual represent an approximate 30 percent sample of the *Fire Fighter I and II Test-Item Bank* developed and maintained by PTS. These testing materials are being used by 47 of the 50 United States, 21 fire service certification agencies worldwide, 121 fire academies, and more than 340 fire department training divisions. For more information on the number of available examination banks and the processes of validation, visit *firetestbanks.com*.

Introduction to the Systematic Approach to Examination Preparation

How does SAEP work? SAEP is an organized process of carefully researched phases that permits each person to proceed through examination preparation at that individual's own pace. At certain points, self-study is required to move from one phase of the program to another. Receiving and then using feedback on one's progress is the basis of SAEP. It is important to follow the program steps carefully to realize the full benefits of the system.

SAEP allows you to prepare carefully for your next comprehensive training, promotional, or certification examination. Just follow the steps to success. PTS, the leader in producing promotional and certification examinations for the Fire and Emergency Medical Service industry for more than 22 years, has both the experience and the testing expertise to help you meet your professional goals.

Using the *Exam Prep* manual will enable you to pinpoint areas of weakness in terms of NFPA 1001, and the feedback will provide the references and page numbers to help you research the questions that you miss or guess using current technical reference materials. This program comprises a three-examination set for Fire Fighter I and Fire Fighter II as described in NFPA 1001, *Standard for Fire Fighter Professional Qualifications*, 2008 Edition.

Primary benefits of SAEP in preparing for these examinations include the following:

- Emphasis on areas of weakness
- Immediate feedback
- Savings in time and energy
- Learning technical material through context and association
- Helpful examination preparation practices and hints

Phases of SAEP

SAEP is organized in four distinct phases for Fire Fighter I and II. These phases are briefly described as follows.

Phase I

Phase I includes three examinations containing items that are selected from each major part of NFPA 1001. An essential part of the SAEP design is to survey your present level of knowledge and build on it for subsequent examination and self-directed study activities. Therefore, it is suggested that you read the reference materials but do not study or look up any answers while taking the initial examination. Upon completion of the initial examination, you will complete a feedback activity and record examination items that you missed or that you guessed. We ask you to perform certain tasks during the feedback activity. Once you have completed the initial examination and researched the answers for any questions you missed, you may proceed to the next examination. This process is repeated through and including the third examination in the Fire Fighter I and II series, depending on the level of certification you are seeking.

Phase II

Fire Fighter II examinations are provided for use in Phase II of SAEP. This phase includes three examinations, each made up of examination items from appropriate sections of NFPA 1001, 2008 Edition.

The examinations should be completed as prescribed in the directions supplied with the examination. Complete the feedback report using the procedures provided in the answer and feedback section. Pay particular attention to those references covering material on which you score the lowest. At this point, it is important to read the materials containing the correct response in context once again. This technique will help you master the material, relate it to other important information, and retain knowledge.

Phase III

Phase III contains important information about examination-item construction. It provides insight regarding the examination-item developers, the way in which they apply their technology, and hints and tips to help you score higher on any examination. Make sure you read this phase carefully. It is a good practice to read it twice, and study the information a day or two prior to your scheduled examination.

Phase IV

Phase IV information addresses the mental and physical aspects of examination preparation. By all means, do not skip this part of your preparation. Points can be lost if you are not ready—both physically and mentally—for the examination. If you have participated in sporting or other competitive events, you know the importance of this level of preparation. There is no substitute for readiness. Just being able to answer the questions will not help you achieve a level of excellence and move you to the top of the examination list for training, promotion, or certification. Quality preparation involves much more than simply answering examination items.

Supplemental Practice Examination Program

The supplemental practice examination program differs from the SAEP program in several ways. First, it is provided over the Internet 24 hours a day, 7 days a week. In addition, this supplemental practice examination allows you to make final preparations immediately before your examination date. You will receive an immediate feedback report that includes the questions missed and the references and page numbers pertaining to those missed questions. The practice examination can help you concentrate on your areas of greatest weakness and will save you time and energy immediately before the examination date. If you choose this method of preparation, do not "cram" for the examination. The upcoming helpful hints for examination preparation will explain the reasons for avoiding a "cramming exercise." A supplemental practice examination is available with the purchase of this *Exam Prep* manual by using the enclosed registration form. Do not forget to fax a copy of your Personal Progress Plotter along with your registration form on page 273. The data supplied on your Personal Progress Plotter will be kept confidential and will be used by PTS to make future improvements in the *Exam Prep* series. You may take a short practice examination to get the procedure clear in your mind by going to *www.webtesting.cc*.

Good luck in your efforts to improve your knowledge and skills. Our primary goal is to improve the Fire and Emergency Medical Service one person at a time. We want your feedback and impressions of the system to help us implement improvements in future editions of the *Exam Prep* series of books. Address your comments and suggestions to *www.firetestbanks.com*.

—————— **Rule 1** ——————

Examination preparation is not easy. Preparation is 95% perspiration and 5% inspiration.

—————— **Rule 2** ——————

Follow the steps very carefully. Do not try to reinvent or shortcut the system. It really works just as it was designed to!

Personal Progress Plotter

Fire Fighter I Exam Prep

Name: _____

Date Started: _____

Date Completed: _____

Fire Fighter II Exam Prep

Name: _____

Date Started: _____

Date Completed: _____

Fire Fighter I	Number Guessed	Number Missed	Examination Score
Examination I-1			
Examination I-2			
Examination I-3			

Fire Fighter II	Number Guessed	Number Missed	Examination Score
Examination II-1			
Examination II-2			
Examination II-3			

Formula to compute Examination Score = ((Number guessed + Number missed) × Point Value per examination item) subtracted from 100.

Note: 100-Item Examination = 1.00 point per examination item
150-Item Examination = .67 points per examination item

Example: In Examination I-3, 5 examination items were guessed and 8 were missed for a total of 13 on a 150-item examination. The examination score would be 100 − (13 × .67 points) = 91.29

Example: Examination II-1, 5 examination items were guessed and 8 were missed for a total of 13 on a 100-item examination. The examination score would be 100 − (13 × 1.00 points) = 87

Note: To receive your free online practice examination, you must fax a copy of your completed Personal Progress Plotter along with your registration form.

PHASE I

Fire Fighter I

Examination I-1, Beginning NFPA 1001

Examination I-1 contains 150 examination items. Read the reference materials but do not study prior to taking the examination. The examination is designed to identify your weakest areas in terms of NFPA 1001. Some steps in SAEP will require self-study of specific reference materials. Remove Examination I-1 from the book. Mark all answers in ink to ensure that no changes are made later. Do not mark through answers or change answers in any way once you have selected your answers.

Step 1—Take Examination I-1. When you have completed Examination I-1, go to Appendix A and compare your answers with the correct answers. Each answer identifies reference materials with the relevant page numbers. If you missed the answer to the examination item, you have a source for conducting your correct answer research.

Step 2—Score Examination I-1. How many examination items did you miss? Write the number of missed examination items in the blank in ink _____. Enter the number of examination items you guessed in this blank _____. Enter these numbers in the designated locations on your Personal Progress Plotter.

Step 3—Now the learning begins! Carefully research the page cited in the reference material for the correct answer. For instance, use Jones and Bartlett, NFPA, *Fundamentals of Fire Fighting Skills, Second Edition*, go to the page number provided, and find the answer.

--------- Rule 3 ---------

Mark with an "X" any examination items for which you guessed the answer. For maximum return on effort, you should also research any answer that you guessed even if you guessed correctly. Find the correct answer, highlight it, and then read the entire paragraph that contains the answer. Be honest and mark all questions on which you guessed. Some examinations have a correction for guessing built into the scoring process. The correction for guessing can reduce your final examination score. If you are guessing, you are not mastering the material.

--------- Rule 4 ---------

Read questions twice if you have any misunderstanding, especially if the question contains complex directions or activities.

─────── **Helpful Hint** ───────

Most of the time your first impression is the best. More than 41% of changed answers during our SAEP field test were changed from a right answer to a wrong answer. Another 33% were changed from one wrong answer to another wrong answer. Only 26% of changed answers were changed from wrong to right. In fact, three participants did not make a perfect score of 100% because they changed one right answer to a wrong one! Think twice before you change your answer. The odds are not in your favor.

─────── **Helpful Hint** ───────

Researching correct answers is one of the most important activities in SAEP. Locate the correct answer for all missed examination items. Highlight the correct answer. Then read the entire paragraph containing the answer. This will put the answer in context for you and provide important learning by association.

─────── **Helpful Hint** ───────

Proceed through all missed examination items using the same technique. Reading the entire paragraph improves retention of the information and helps you develop an association with the material and learn the correct answers. This step may sound simple. A major finding during the development and field testing of SAEP was that you learn from your mistakes.

Examination I-1

Directions

Remove Examination I-1 from the manual. First, take a careful look at the examination. There should be 150 examination items. Notice that a blank line precedes each examination item number. This line is provided for you to enter the answer to the examination item. Write the answer in ink. Remember the rule about not changing your answers. Our research shows that changed answers are most often changed to an incorrect answer, and, more often than not, the answer that is chosen first is correct.

If you guess the answer to a question, place an "X" or a check mark by your answer. This step is vitally important as you gain and master knowledge. We will explain how we treat the "guessed" items later in SAEP.

Take the examination. Once you complete it, go to Appendix A and score your examination. Once the examination is scored, carefully follow the directions for feedback on the missed and guessed examination items.

_____ **1.** NFPA 1001 is the:
   A. Standard for Fire Fighter Professional Qualifications.
   B. Standard for Industrial Fire Brigades.
   C. Standard for Fire Department Occupational Safety and Health Program.
   D. Standard for Blood Borne Pathogen Training.

_____ **2.** One of the **primary** functions of the truck/ladder company is:
   A. performing forcible entry to fire building.
   B. directing traffic at fire scene.
   C. laying supply lines for engine companies.
   D. fire attack.

_____ **3.** OSHA may establish regulations governing fire department activities such as:
   A. work hours.
   B. anti-discrimination laws.
   C. the level of experience required to hold a chief's office.
   D. health and safety.

_____ **4.** In the Incident Command System (ICS), the functional area responsible for all incident activities, including the development and implementation of strategic decisions, is:
   A. Planning.
   B. Operations.
   C. Logistics.
   D. Command.

_____ **5.** What normal body cooling mechanism is lost when wearing PPE?
   A. Radiation from the head
   B. Radiation from the body
   C. Internal cooling from ingestion
   D. Evaporation of perspiration

_____ **6.** What is one aspect of high-rise fires that is especially draining of energy?
    A. Walking up many flights of stairs in PPE
    B. Taking the elevator up only to the floor below the fire
    C. The inevitable compromise of the HVAC system
    D. Establishing the staging sometimes many floors below the fire floor

_____ **7.** A _____ permits fire fighters to gain access during an emergency.
    A. lockout
    B. tagout
    C. Baker box
    D. key box/lock box

_____ **8.** The _____ is tied as illustrated below:
    A. bowline
    B. becket/sheet bend
    C. clove hitch
    D. figure eight

_____ **9.** In kernmantle rope, the kern or core of the rope accounts for approximately
_____ of its strength.
    A. 50 percent
    B. 90 percent
    C. 75 percent
    D. 25 percent

_____ **10.** The _____ is tied as illustrated in the drawing below.
   A. bowline
   B. half hitch
   C. clove hitch
   D. figure-eight

_____ **11.** The combination of knots recommended to hoist a pike pole includes:
   A. a becket/sheet bend with a bight.
   B. several half hitches.
   C. bowline and half hitches.
   D. a clove hitch and half hitches.

_____ **12.** The order of rank and authority in the fire service describes:
   A. chain of command.
   B. discipline.
   C. span of control.
   D. unity of command.

_____ **13.** The street, or address side of the structure is termed:
   A. Division A.
   B. Division B.
   C. Division C.
   D. Division D.

_____ **14.** The ratio of the mass of a given volume of liquid compared with the mass of an equal volume of water at the same temperature is:
   A. specific gravity.
   B. vapor density.
   C. flash point.
   D. surface to mass ratio.

_____ **15.** Life-safety rope should be stored in:
   A. coils or in rope bags.
   B. a large bundle.
   C. moisture-proof cases.
   D. coiled in mesh bags.

_____ **16.** Toxic atmospheres due to smoke and heat are called:
   A. fire fighter hazardous environments.
   B. immediately dangerous to life and health.
   C. carcinogenic hazard areas.
   D. exclusionary zones.

_____ **17.** Which of the following has the highest priority for a radio transmission?
   A. Notification from dispatch of road closures
   B. Emergency traffic from a unit working at a fire or rescue
   C. Vital signs of a patient being transported by fire department ambulance
   D. Transmission of local forest fire danger notice

_____ **18.** The telecommunicator's **first** responsibility is to:
   A. obtain the information that is required to dispatch the appropriate units to the correct location.
   B. reassure the caller that help will be forthcoming shortly and that his/her problem can be dealt with.
   C. provide the caller with real-time, emergency instructions such as how to perform CPR.
   D. ascertain the exact type of event that the caller is reporting.

_____ **19.** Before transmitting any information over the fire department radio, a fire fighter should:
   A. press the key two or three times to signal the intent to transmit.
   B. key the microphone and then clear his/her throat to be sure his/her voice will be clear.
   C. turn up the volume control if his/her voice is naturally soft.
   D. listen to be sure the channel is **not** being used.

_____ **20.** When using positive pressure SCBA, a poor seal between the face piece and the fire fighter's face is:
   A. **not** dangerous, because the positive pressure will keep toxic gases out of the face piece.
   B. dangerous, because it is depleting the air supply more quickly.
   C. **not** possible, because the positive pressure will seal the face piece to the face.
   D. the main cause of deaths on the fireground.

_____ **21.** The bypass valve on a self-contained breathing apparatus is used:
   A. during hazardous materials incidents.
   B. in emergency situations involving a malfunctioning regulator.
   C. to clear the mask of unwanted condensation.
   D. to cool the face piece when high heat is encountered.

_____ **22.** The **first** noticeable signs of oxygen deficiency are:
   A. profuse sweating and ringing in the ears.
   B. dizziness, impaired vision, and giddiness.
   C. increased respiratory rate and impaired muscular coordination.
   D. headache and rapid fatigue.

_____ **23.** Four hazardous atmospheres that fire fighters are likely to encounter at a fire are:
    A. super-heated air, toxic gases, oxygen deficiency, and smoke.
    B. toxic gases, hyperventilation, oxygen deficiency, and heat.
    C. heat, light, smoke, and chemical chain reaction.
    D. carbon monoxide, ammonia, water, and hydrogen sulfide.

_____ **24.** Trapped fire fighters awaiting rescue will use less air if they:
    A. partially close the cylinder valve.
    B. open the bypass valve.
    C. struggle to get free.
    D. control their breathing.

_____ **25.** The **primary** function of the bypass valve on SCBA is:
    A. to enable the wearer to breathe more oxygen.
    B. for use if the regulator fails.
    C. to help control excessive heat in the face piece.
    D. to facilitate removal of condensation from the lens of the face piece.

_____ **26.** The proper position of the bypass valve on positive-pressure SCBA under normal conditions is:
    A. fully open.
    B. cracked open.
    C. fully closed.
    D. open three full turns.

_____ **27.** Hypoxia is:
    A. low blood sugar.
    B. low blood pressure.
    C. a deficiency of oxygen.
    D. a lowering of the body's core temperature.

_____ **28.** Inhaled toxic gases can directly cause:
    A. disease of the lung tissue.
    B. muscle cramps in the lower extremities.
    C. blurred vision, leading to blindness.
    D. amnesia.

_____ **29.** Which of the following statements regarding the two SCBA types in fire service use is **incorrect**?
    A. In an open-circuit SCBA, exhaled air is vented to the outside atmosphere.
    B. In a closed-circuit SCBA, exhaled air stays in the system for filtering, cleaning, and circulation.
    C. The closed-circuit type is commonly used for structural firefighting.
    D. SCBAs for fire service use are designed and built in accordance with NIOSH and NFPA standards.

_____ **30.** What is the <u>**first**</u> <u>**step**</u> a fire fighter should perform before donning the SCBA?
   A. Check the air supply of the SCBA unit.
   B. Check the PASS device.
   C. Test the low air alarm.
   D. Loosen all straps on the harness.

_____ **31.** The purpose of a pass device is to:
   A. sound an alarm if certain fire or asphyxiate gases are detected.
   B. warn fire fighters when their air supply is low.
   C. sound an alarm if a fire fighter is motionless for a set period of time.
   D. keep track of elapsed time in deteriorating conditions.

_____ **32.** Oxygen deficient atmospheres are those having less
   than _____ percent oxygen in the air.
   A. 21
   B. 20.5
   C. 20
   D. 19.5

_____ **33.** At the beginning of the shift, the fire fighter should make sure their SCBA is at
   least _____ percent full.
   A. 75
   B. 80
   C. 85
   D. 90

_____ **34.** There are two methods that can be used to don SCBA that is stored in a case:
   A. over-the-head and coat method.
   B. compartment and coat method.
   C. over-the-head and compartment method.
   D. jacket and backup mount.

_____ **35.** Manufacturers should provide users of PPE with which of the
   following information?
   A. Cleaning instructions
   B. MSDS information
   C. Shelf life
   D. Liability protection

_____ **36.** The purpose for the use of reflective trim on PPE is to:
   A. increase the visibility of the wearer to others.
   B. provide protection for material under the trim.
   C. allow the wearer to blend in with the surroundings.
   D. be more stylish than the plain PPE.

_____ **37.** Tempered plate glass should be broken only as a last resort. It is recommended that it be shattered by striking:
   A. in the center with a large, blunt object.
   B. at the lowest corner with the pick end of a fire ax.
   C. with a flat-headed ax with hands above the head.
   D. with a flat-headed ax at the highest corner.

_____ **38.** In what type of occupancy will fire fighters usually find sliding doors?
   A. Barns or warehouses
   B. Commercial
   C. Residential
   D. Institutional

_____ **39.** Once overhead doors have been forced, they should be:
   A. removed.
   B. unlocked to prevent locking.
   C. locked.
   D. blocked open.

_____ **40.** A K-Tool is correctly positioned when the blades bite into the:
   A. cylinder.
   B. outer rim.
   C. face.
   D. keeper.

_____ **41.** A/An _____ can be safely carried by two fire fighters, although it is designed to be used by two or four fire fighters.
   A. hux bar
   B. Halligan tool
   C. battering ram
   D. oxyacetylene cutting unit

_____ **42.** Which of the following tools is considered to be a pulling tool?
   A. Crowbar
   B. Flat-head ax
   C. Pike pole
   D. Halligan tool

_____ **43.** "Irons" are formed by combining a:
   A. Halligan and a crowbar.
   B. pickhead ax and a crowbar.
   C. pickhead ax and a maul.
   D. flathead ax and a Halligan.

_____ **44.** What is one example of a hand-powered hydraulic spreader tool?
   A. Spanner tool
   B. Pompier tool
   C. Rabbet tool
   D. Pitot tool

_____ **45.** Which of the following statements regarding power saw safety is <u>incorrect</u>?
    A. Use the right blade for the material being cut.
    B. The saw should be started on level ground and carried up the ladder while running.
    C. Power saws require two fire fighters—the saw operator and a guide fire fighter.
    D. Conduct daily checks for operation and blade condition.

_____ **46.** Which tool is used to pull out a cylinder lock mounted in a wood or heavy metal door?
    A. Rabbet tool
    B. Hux bar
    C. K-tool
    D. Ax

_____ **47.** The most important factor to consider in forcible entry is to:
    A. select the right tool to do the job.
    B. use the tool that can do the job fastest.
    C. use prying tools before striking tools.
    D. use striking tools before prying tools.

_____ **48.** Before fire fighters enter a burning building to perform rescue work, they must first consider:
    A. manpower on the scene.
    B. weather conditions.
    C. damaging evidence of forced entry.
    D. their own safety.

_____ **49.** The term for a roll call taken by a supervisor at an emergency incident is a:
    A. Fire Fighter Location Confirmation.
    B. Time and Location Verification.
    C. System Communications Check.
    D. Personnel Accountability Report.

_____ **50.** OSHA regulations require that when fire fighters are inside a structure fire, at least _____ trained and equipped fire fighters must be standing by outside.
    A. 2
    B. 3
    C. 4
    D. 6

_____ **51.** When referring to ladders, the term tie rod is used to describe a metal rod running from:
    A. one beam to the other.
    B. a beam to a rung.
    C. rung to truss block.
    D. the pulley(s) to the ladder frame.

_____ **52.** When a ladder is raised, it should be placed at an angle of
approximately _____ to ensure a safe climb.
  A. 55°
  B. 75°
  C. 65°
  D. 45°

_____ **53.** A nonadjustable ladder that consists of only one section is
called a _____ ladder.
  A. combination
  B. single/wall
  C. folding
  D. special-use

_____ **54.** Ladders should extend five rungs above the roof edge to aid fire fighters in
climbing onto and off the ladder and:
  A. so fire fighters can find the ladder in heavy smoke conditions.
  B. to help locate victims who may be trapped.
  C. to prevent damage to the tips of the beams.
  D. to keep the ladder cool under heavy fire conditions.

_____ **55.** A fire fighter who is climbing a ladder and carrying a tool in one hand, should:
  A. grasp only every other rung during the climb.
  B. place the tool at every other rung during the climb for balance.
  C. slide the free hand up the underside of the beam while making the climb.
  D. never climb a ladder with only one hand free.

_____ **56.** When performing the one-person raise, the _____ of the ladder
is placed against the building wall.
  A. tip
  B. beam
  C. butt
  D. stay poles

_____ **57.** The common length of the straight ground ladder is usually
between _____ feet.
  A. 6 to 14
  B. 12 to 24
  C. 18 to 30
  D. 25 to 50

_____ **58.** What are two methods of raising a ladder to the vertical position?
  A. One-person raise and two-person raise
  B. Manual raise and mechanical raise
  C. East raise and West raise
  D. Rung raise and beam raise

_____ **59.** What are butt spurs on a ladder?
    A. Protrusions that attach the hose strap to the ladder
    B. The teeth of the cam system that extends the fly sections
    C. Bumps and dents on the foot pad
    D. Spikes on the base that keep it from slipping

_____ **60.** What do the stops do on an extension ladder?
    A. Prevent it from over-extending
    B. Limit how far the fly sections can collapse
    C. Lock the extensions in place when extended
    D. Prevent the whole ladder from slipping on the ground

_____ **61.** Where does the top beam of the ladder rest in the one-fire fighter carry?
    A. On the hip
    B. In the hand
    C. On the shoulder
    D. On the crook of the elbow

_____ **62.** One of the most useful tools to aid in handling a charged hose line is a hose:
    A. wrench.
    B. jacket.
    C. strap.
    D. clamp.

_____ **63.** One-inch rubber-covered and rubber-lined hose equipped with one-inch couplings is commonly called a:
    A. forestry hose.
    B. supply hose.
    C. booster hose.
    D. engine line.

_____ **64.** In the fire service, the basic definition of the word "rescue" is:
    A. removing a victim from a hazardous situation to safety.
    B. stabilizing a victim before transporting.
    C. performing cardiopulmonary resuscitation (CPR) on a victim.
    D. All of the above.

_____ **65.** During search operations, the fire fighter should:
    A. rest to conserve air.
    B. regroup with all those searching.
    C. listen for calls or signals for help.
    D. leave a trail to aid escape.

_____ **66.** Which of the following are characteristics of a **primary** search?
    A. Rapid and systematic
    B. Slow and deliberate
    C. Accomplished only with a charged hose line
    D. Begun only after top-side ventilation is provided

_____ **67.** When conducting a **primary** search within a structure, a fire fighter should begin:

    A. in the center of the room.

    B. on a wall.

    C. always start with right hand pattern.

    D. under or behind furnishings.

_____ **68.** During a search of a building involved in fire, if a fire fighter becomes disoriented, the fire fighter should **attempt** to:

    A. remain calm.

    B. retrace steps to original location.

    C. seek a place of refuge and activate PASS device.

    D. All of the above.

_____ **69.** Which of the following statements is **incorrect**?

    A. The secondary search is the most dangerous.

    B. Searching a building is completed in two different operations; primary and secondary search.

    C. During primary search, the team is often ahead of attack lines and may be above the fire.

    D. The primary search takes place in a rapid but thorough manner in areas most likely to have victims.

_____ **70.** A rapid intervention crew team is composed of:

    A. at least four fire fighters.

    B. fire fighters to rescue occupants if found.

    C. at least two fire fighters.

    D. fire fighters waiting by ready to don PPE if necessary.

_____ **71.** Immobilizing a victim who is suspected of having a spinal injury on a long backboard requires _____ rescuers.

    A. 2

    B. 3

    C. 4

    D. 5

_____ **72.** When exposed to products of combustion, the _____ are more vulnerable to injury than any other body area.

    A. lungs and respiratory tract

    B. heart and respiratory tract

    C. lungs and eyes

    D. brain and spinal cord

_____ **73.** What does the E stand for in Lower _____ Limit (LEL)?

    A. Endothermic

    B. Energetic

    C. Explosive

    D. Evolution

_____ **74.** In a multi-story building with a standpipe system, a fire fighter should make the connection _____ the fire floor.
    A. on the floor below
    B. on the floor above
    C. on
    D. two levels above

_____ **75.** Fine water droplets and maximum high water surface area are characteristics of a _____ stream.
    A. solid
    B. fog
    C. broken
    D. straight

_____ **76.** A stream designed to be as compact as possible with little shower or spray is known as a _____ stream.
    A. solid
    B. fog
    C. straight
    D. narrow-angle fog

_____ **77.** When advancing a <u>dry</u> hose line up a ladder, fire fighters should position themselves on the ladder:
    A. on opposing sides of the beam.
    B. no more than six feet apart.
    C. with no more than one fire fighter per section.
    D. within arms' reach of each other.

_____ **78.** A hose _____ is used to seal small cuts or breaks that may occur in fire hose or to connect mismatched or damaged couplings of the same size to stop leaking.
    A. bridge
    B. clamp
    C. jacket
    D. seal

_____ **79.** A straight stream is discharged from a/an _____ nozzle.
    A. smooth-orifice
    B. adjustable-fog
    C. special-purpose
    D. applicator

_____ **80.** The _____ method is performed with <u>two</u> fire fighters and used for breaking tight couplings without the use of a spanner wrench.
    A. stiff-arm
    B. foot-tilt
    C. knee-press
    D. coupling-tilt

_____ **81.** Under what circumstances should master stream devices be directed into buildings where fire fighters are operating inside?
  A. Only when the whereabouts of every person inside is known to the master stream operator
  B. Only when the fire is clearly getting the better of the inside crew
  C. Under no circumstances
  D. Under any circumstances

_____ **82.** A dutchman is:
  A. used when loading hose and has a short fold or a reverse fold with a coupling at the point where a fold should occur.
  B. a process for uncoupling hose.
  C. a short length of hose that connects a deluge set to the engine company's pump.
  D. the fold in a twin donut roll.

_____ **83.** When should the fire fighter attaching hose to the hydrant charge the supply line?
  A. As soon as possible
  B. Not until the driver/operator's signal is received
  C. After slowly opening the hydrant all the way
  D. After ten full minutes from the time of drop-off

_____ **84.** Which of the following statements regarding fog nozzles **is** **not** correct?
  A. Adjustable fog nozzles provide personal protection because of the screening effect between them and the fire.
  B. Fog nozzles provide better heat absorption.
  C. They have a greater reach than solid stream nozzles.
  D. Fog streams can produce more steam that can extinguish hidden fire.

_____ **85.** Which of the following is a recommended practice to prevent mechanical fire hose damage?
  A. Maintain the same fold positions when reloading hose on apparatus.
  B. Eliminate all chafing blocks in the vicinity of the operating pumper.
  C. Open and close nozzles, valves and hydrants slowly.
  D. Avoid inserting hose in ramps or bridges.

_____ **86.** Proper ventilation results in an orderly movement of _____ through and out of the structure.
  A. water fog
  B. hose line crews
  C. rescue personnel
  D. heated fire gases

_____ **87.** Convection is:
  A. transfer of heat through space by infrared rays.
  B. transfer of heat through a solid medium.
  C. not considered a method of heat transfer.
  D. transfer of heat through liquids or gases by circulating currents.

_____ **88.** Heat transfer that takes place in the form of electromagnetic waves is known as:
A. convection.
B. radiation.
C. conduction.
D. flame contact.

_____ **89.** Which stage of fire releases the maximum amount of heat for the available fuel and produces large volumes of fire gases?
A. Ignition
B. Fully developed
C. Growth
D. Rapid oxidation

_____ **90.** Which of the following is the first stage/phase of fire?
A. Incipient/ignition
B. Fully developed
C. Growth
D. Decay

_____ **91.** The term vapor density refers to the weight of a gas as compared to the weight of:
A. water.
B. air.
C. carbon.
D. nitrogen.

_____ **92.** If a gas has a vapor density greater than one when it escapes from its container:
A. it will rise.
B. its movement will be dependent on wind direction and speed.
C. its movement will be dependent on temperature.
D. it will sink and collect at low points.

_____ **93.** The acronym BLEVE stands for:
A. barometric level emergency valve enclosure.
B. bring local emergency vehicles early.
C. boiling liquid expanding vapor explosion.
D. boiling liquid emergency valve exit.

_____ **94.** Positive-pressure ventilation is:
A. pulling smoke out.
B. blowing fresh air in.
C. removing oxygen.
D. increasing thermal layering.

_____ **95.** From a life safety point of view, the advantage(s) of proper ventilation for building occupants is/are that it:
A. improves visibility.
B. reduces the danger of backdraft explosions.
C. removes toxic smoke.
D. All of the above.

_____ **96.** Smoke and heat collects at a structure starting from the:
   A. lowest point.
   B. windward side.
   C. highest point.
   D. leeward side.

_____ **97.** _____ construction is often used in mill construction.
   A. Frame
   B. Ordinary
   C. Heavy timber
   D. Noncombustible

_____ **98.** The usual cause of collapse of open web steel joist is the:
   A. amount of heat generated by the fire in a structure.
   B. poor method of construction.
   C. impact load of fire fighters on the roof.
   D. All of the above.

_____ **99.** Which of the following roof structures provide(s) quick and effective
   initial ventilation?
   A. Scuttle covers
   B. Roof level skylights
   C. Hatches
   D. All of the above.

_____**100.** <u>Modern</u> wood-frame construction uses a technique that builds one floor at a
   time and inserts a plate between each floor that acts as a fire stop. This
   technique is called:
   A. flitch-plate.
   B. platform-frame.
   C. balustrade.
   D. awning.

_____ **101.** What does vertical ventilation require that horizontal ventilation does <u>not</u>?
   A. Positive-pressure fans
   B. Openings in the roof or highest part of the building
   C. Negative-pressure fans
   D. Airtightness everywhere but the vent opening

_____**102.** When trapped moisture in concrete is heated to steam and expands, causing
   parts of the concrete to break away, this is known as:
   A. delamination.
   B. depolymerizing.
   C. spalling.
   D. spation.

_____**103.** Searching for hidden fires is a **primary** function of:
   A. sizeup.
   B. salvage.
   C. fire investigation.
   D. overhaul.

_____**104.** The search for and extinguishment of hidden fire and placing the building in a safe condition is known as:
   A. overhaul.
   B. secondary search.
   C. sizeup.
   D. salvage.

_____**105.** Overhaul operations should commence:
   A. at the same time as suppression.
   B. after fire investigation concerns are met.
   C. immediately after fire suppression.
   D. prior to the first line going into operation.

_____**106.** Ideally, when should fire investigators begin their work at a fire scene?
   A. After overhaul is complete
   B. Before overhaul starts
   C. Any time other than when overhaul operations are occurring
   D. During salvage operations

_____**107.** One way to remove water coming through the ceiling from upper floors is by the use of:
   A. sponges.
   B. chutes.
   C. carryalls.
   D. floor runners.

_____**108.** If large amounts of water need to be removed from areas lacking floor drains, fire fighters may _____ to remove the water quickly.
   A. drill holes in the floor
   B. remove toilets and use the sewer system as a drain
   C. drill holes in exterior walls
   D. open exterior walls to allow for sufficient drainage

_____**109.** Which of the following statements is **incorrect**?
   A. A catch-all is used when the volume of water is too great for a water chute.
   B. The balloon toss requires two fire fighters.
   C. A water chute is used when the volume of water is too great for a catch-all.
   D. The shoulder toss is a one-person operation.

_____ **110.** When the water flow alarm (water gong) sounds, this indicates that:
   A. water has stopped flowing in the system.
   B. heat detection devices have been activated and one may expect the deluge set to begin discharging water momentarily.
   C. water is flowing in the system.
   D. a heat actuating device has been activated and someone should turn the main sprinkler valve to the open position.

_____ **111.** The flow of water from an individual sprinkler head can be controlled by using:
   A. the main system control valve only.
   B. sprinkler wedges or tongs.
   C. the valve required at each branch line.
   D. individual controls.

_____ **112.** Sprinkler heads rated for the Ultra High temperature classification are color coded:
   A. yellow.
   B. white.
   C. blue.
   D. orange.

_____ **113.** During salvage, what should be done with any pictures on the walls?
   A. They should be left in place as they are.
   B. They should be left in place, but turned toward the wall.
   C. They should be removed and placed with the furniture.
   D. They should be set on the floor against the wall, facing the wall.

_____ **114.** One limitation of backpack water vacuums is that they:
   A. cannot pull water up from the floor.
   B. cannot be worn with SCBA.
   C. can only be used once.
   D. are illegal in many places.

_____ **115.** Soft sleeve intake hose is used for:
   A. transferring water from a hydrant to an apparatus.
   B. primarily drafting water from a static source.
   C. siphoning water from one portable tank to another.
   D. transferring water from pump to tank.

_____ **116.** Tenders combined with _____ can efficiently provide large volumes of water to a fireground operation.
   A. large diameter hose
   B. automatic nozzles
   C. portable water tanks
   D. ladder trucks

_____ **117.** The two major hydrant types are:
   A. wet-barrel and dry-barrel.
   B. high-pressure and low-pressure.
   C. ground water and surface water.
   D. treated water and untreated water.

_____ **118.** Which of the following statements regarding dry hydrants is **<u>incorrect</u>**?
   A. They are used primarily in rural areas with no water systems.
   B. They are a pipe system with a pumper suction connection at one end and a strainer at the other.
   C. Dry hydrant is another term for dry barrel hydrant.
   D. They are a connection point for drafting from a static water source.

_____ **119.** Residual pressure is:
   A. the pressure in the system with no hydrants or water flowing.
   B. the pressure remaining in a system after water has begun flowing.
   C. the level of ground water under the surface.
   D. a device that speeds the unloading of water from a tender.

_____ **120.** On a dry-barrel hydrant, the valve that controls water flow to all outlets is located _____ of the hydrant.
   A. at the top
   B. on the side
   C. at the base
   D. by each outlet

_____ **121.** All portable extinguishers are classified according to their:
   A. size.
   B. freeze potential.
   C. intended use.
   D. conductivity.

_____ **122.** A dry chemical extinguisher rated 60-B is capable of extinguishing a _____ flammable liquid pan fire.
   A. 40 ft²
   B. 60 ft²
   C. 120 ft²
   D. 150 ft²

_____ **123.** Extinguishing agents safe for use on fires in or near energized electrical equipment include:
   A. dry powder, carbon dioxide, and AFFF.
   B. carbon dioxide and dry chemical.
   C. dry chemical, pressurized foam, and carbon dioxide.
   D. AFFF, carbon dioxide, and dry chemical.

_____**124.** A green triangle containing a letter would indicate an extinguisher to be used
on _____ fires.
A. Class A
B. Class B
C. Class C
D. Class D

_____**125.** Extinguishers suitable for Class A fires can be identified by
a _____ containing the letter "A." If colored, it
should be _____.
A. circle, blue
B. star, yellow
C. triangle, green
D. square, red

_____**126.** Extinguishers suitable for Class C fires can be identified by
a _____ containing the letter C.
A. yellow star
B. green triangle
C. red square
D. blue circle

_____**127.** Extinguishers suitable for Class D fires can be identified by
a _____ containing the letter D.
A. blue circle
B. yellow star
C. green triangle
D. red square

_____**128.** Fires involving combustible metals such as magnesium, titanium, zirconium,
sodium, and potassium, are _____ fires.
A. Class A
B. Class B
C. Class C
D. Class D

_____**129.** A carbon dioxide (CO_2) extinguisher's means of discharge is:
A. chemical reaction.
B. stored liquefied compressed gas.
C. cartridge activation.
D. manual hand-pump.

_____**130.** A pump tank extinguisher rated as 4-A can be expected to extinguish
approximately _____ as much fire as one rated 2-A.
A. twice
B. three times
C. four times
D. eight times

_____ **131.** A dry chemical fire extinguisher rated as 10-B should be capable of extinguishing _____ times as much fire as a unit rated as 1-B.
A. 5
B. 10
C. 20
D. 100

_____**132.** Fire extinguisher classification symbols are displayed by all of the following **except**:
A. color.
B. shape.
C. letter.
D. weight of container.

_____**133.** What injury could occur by placing your hand on the horn of a CO_2 extinguisher while discharging it?
A. Cold can freeze the skin.
B. Heat from friction can cause burns.
C. Electrical shock could occur.
D. The horn could be distorted leading to poor application.

_____**134.** Class K fires involve:
A. atomic material.
B. computer network equipment.
C. hazardous waste.
D. high temperature cooking oils, such as vegetable or animal oils and fats.

_____**135.** Cold temperatures would have the **greatest** effect on _____ extinguishing agents stored in an extinguisher.
A. dry powder
B. water-based
C. dry chemical
D. carbon dioxide

_____**136.** Regular dry chemicals (sodium bicarbonate-based and potassium-based) work on _____ fires.
A. only Class D
B. Class A
C. only Class B
D. Classes B and C

_____**137.** The acronym for the four-step process for proper use of fire extinguishers is:
A. PASS.
B. RACE.
C. PACE.
D. DUMP.

_____**138.** What is the term for water-soluble flammable liquids such as alcohols, acetone, and others?
    A. Volatile fuels
    B. Polar solvents
    C. Flammable surfactants
    D. Three-dimensional liquids

_____**139.** Fire fighters should treat all downed wires as:
    A. energized.
    B. safe if in contact with the ground.
    C. only dangerous if nearby homes have power.
    D. safe if <u>not</u> arcing.

_____**140.** Which one of the following statements regarding wildland firefighting is <u>incorrect</u>?
    A. From a flat to a 30° slope, the fire will double its rate of spread.
    B. A ridge fire does <u>not</u> tend to draw fire to itself.
    C. A fire burns downhill faster than it does uphill.
    D. A fire burns uphill faster than it does downhill.

_____**141.** Worn, damaged, and deteriorated parts of a SCBA must be replaced according to:
    A. past practice.
    B. NIOSH/OSHA Respiratory Protection Act.
    C. manufacturer's instructions.
    D. the wearer's recommendations.

_____**142.** Composite SCBA cylinders must be hydrostatically tested every:
    A. year.
    B. three years.
    C. six years.
    D. ten years.

_____**143.** Steel and aluminum cylinders for breathing apparatus should be hydrostatically tested after each _____ year period.
    A. two
    B. three
    C. four
    D. five

_____**144.** When inspecting kernmantle rope for damage, one can assume:
    A. irregularities in shape or weave are of no concern.
    B. the rope's core can be damaged without showing evidence of outer sheath damage.
    C. 75 percent of the rope's strength lies in the outer sheath.
    D. that if the rope fails an inspection, it may still be used as a life safety rope.

_____**145.** To clean or remove any oily or greasy residues from a metal ground ladder, it is recommended to use:
   A. good quality waxing compound.
   B. clear water and an approved detergent.
   C. a top quality commercial lacquer thinner.
   D. sandpaper and steel wool.

_____**146.** When assigned to perform routine maintenance on ground ladders, a fire fighter should do all of the following **except**:
   A. remove dirt with a brush and running water.
   B. use an approved detergent to remove oily residues.
   C. look for structural defects in the ladder.
   D. perform an extension ladder hardware test.

_____**147.** It is permissible to use paint on a fire department ladder:
   A. to mark the bed section on multisection ladders.
   B. when there is a possibility of dry rot.
   C. to mark the ladder ends for visibility.
   D. when salt water may be a problem.

_____**148.** When loading fire hose, the _____ serves to change the direction of the hose and can also change the location of a coupling.
   A. horseshoe
   B. flat load
   C. reverse lay
   D. Dutchman

_____**149.** Which coupling goes on the inside of a straight hose roll?
   A. The female
   B. The male
   C. Either A or B.
   D. It depends on the diameter.

_____**150.** Which of the following **is not** a common load for preconnected attack lines?
   A. Flat load
   B. Triple layer load
   C. Straight load
   D. Minuteman load

_____

Did you score higher than 80 percent on Examination I-1? Circle Yes or No in ink.

Now that you have finished the Feedback Step for Examination I-1, it is time to repeat the process by taking another comprehensive examination of the NFPA 1001 Standard.

Examination I-2, Adding Difficulty and Depth

During Examination I-2, progress will be made in developing your depth of knowledge and skills.

   Step 1—Take Examination I-2. When you have completed Examination I-2, go to Appendix A and compare your answers with the correct answers.

   Step 2—Score Examination I-2. How many examination items did you miss? Write the number of missed examination items in the blank in ink _____. Enter the number of examination items you guessed in this blank _____. Enter these numbers in the designated locations on your Personal Progress Plotter.

   Step 3—Once again, the learning begins. During the feedback step, research the correct answer using the Appendix A information for Examination I-2. Highlight the correct answer during your research of the reference materials. Read the entire paragraph containing the correct answer.

Helpful Hint

Follow each step carefully to realize the best return on effort. Would you consider investing your money in a venture without some chance of earning a return on that investment? Examination preparation is no different. You are investing time and expecting a significant return for that time. If, indeed, time is money, then you are investing money and are due a return on that investment. Doing things right and doing the right things in examination preparation will ensure the maximum return on effort.

Examination I-2

Directions

Remove Examination I-2 from the manual. First, take a careful look at the examination. There should be 150 examination items. Notice that a blank line precedes each examination item number. This line is provided for you to enter the answer to the examination item. Write the answer in ink. Remember the rule about not changing your answers. Our research shows that changed answers are most often changed to an incorrect answer, and, more often than not, the answer that is chosen first is correct.

If you guess an answer, place an "X" or a check mark by your answer. This step is vitally important to gain and master knowledge. We will explain how we treat the "guessed" items later in SAEP.

Take the examination. Once you complete it, go to Appendix A and score your examination. After the examination is scored, carefully follow the directions for feedback of the missed and guessed examination items.

_____ **1.** Life safety, incident stabilization, and _____ are the three most important organizational duties for fire departments to pursue.
   A. building inspections
   B. public information
   C. property conservation
   D. resource management

_____ **2.** The plan or written document for tactical operations is known as a department's:
   A. S.O.P./S.O.G.
   B. organizational chart.
   C. prefire plan.
   D. mission statement.

_____ **3.** Fire department standard operating procedures/standard operating guides should be established in the **most** **commonly** accepted order of fireground priorities, which is:
   A. life safety, property conservation, fire control.
   B. property conservation, life safety, fire control.
   C. fire control, life safety, property conservation.
   D. life safety, incident stabilization, and property conservation.

_____ **4.** In the Incident Command System (ICS), the functional area responsible for providing facilities, services, and materials necessary to support an incident is:
   A. Planning.
   B. Operations.
   C. Logistics.
   D. Command.

_____ **5.** Historically, the basic unit of a fire department is:
   A. the fire fighter.
   B. the company.
   C. a battalion.
   D. a platoon/shift.

_____ **6.** In the Incident Command System (ICS), the functional area that directs the organization's tactics to meet the strategic goals developed by command and is responsible for the management of all activities applicable to the primary mission is:
   A. Planning.
   B. Operations.
   C. Logistics.
   D. Command.

_____ **7.** _____ relates to the number of personnel an individual can effectively supervise.
   A. Staff rule
   B. Span of control
   C. Rule of thumb
   D. Line rule

_____ **8.** One of the major advantages of an Incident Command System (ICS) is that it allows agencies to communicate using common _____ and operating procedures.
   A. terminology
   B. personnel
   C. apparatus
   D. equipment

_____ **9.** What is developed to provide definite guidelines for present and future actions?
   A. Analyses
   B. Policies
   C. Comparisons
   D. Post-incident plans

_____ **10.** What is the principle called that says each fire fighter answers to only one supervisor?
   A. Span of responsibility
   B. Organizational accountability
   C. Unity of command
   D. Delegation of duty

_____ **11.** The **primary** role of a fire alarm system is to:
   A. notify the fire department.
   B. hold the fire in check.
   C. alert occupants.
   D. summon help.

_____ **12.** What is the term for locations from which an engine can draft out of a static water supply source?
   A. Drafting/fill sites
   B. Standpipes
   C. Source locations
   D. Relay origins

_____ **13.** Structural firefighting Personal Protective Equipment is designed to be worn with what specific piece of equipment?
  A. Rappelling harness
  B. SCBA
  C. Flotation vest
  D. Chemical splash suit

_____ **14.** The **<u>primary</u>** use of the _____ is to attach a rope to a round object such as a pike pole.
  A. clove hitch
  B. bowline
  C. becket/sheet bend
  D. rescue knot

_____ **15.** The part of the rope that is used for work such as hoisting or pulling is called the:
  A. working end.
  B. round turn.
  C. running end.
  D. standing part.

_____ **16.** The combination of knots used to hoist a hose line involves:
  A. the bowline-on-bight and half hitch.
  B. the half-sheep shank and half hitch.
  C. a clove hitch and half hitch.
  D. the bowline and the sheep shank.

_____ **17.** Fire service rope falls into two use classifications:
  A. life safety and utility.
  B. braided and kernmantle.
  C. dynamic and static.
  D. natural and synthetic.

_____ **18.** An acceptable use for the _____ is to attach a rope to an object such as a pole, post, or hose line.
  A. water knot
  B. clove hitch
  C. figure-eight
  D. becket/sheet bend

_____ **19.** The knot used to tie two ropes of unequal diameter together is the:
  A. clove hitch.
  B. square knot.
  C. becket/sheet bend.
  D. half hitch.

_____ **20.** The _____ is tied as illustrated in the drawing below.
   A. bowline
   B. becket/sheet bend
   C. clove hitch
   D. figure eight

_____ **21.** The _____ is tied as illustrated in the drawing below.
   A. bowline
   B. becket/sheet bend
   C. square knot
   D. follow through

_____ **22.** The drawing below depicts one of the three elements of a knot or hitch. It is known as a:
   A. bight.
   B. loop.
   C. round turn.
   D. clove hitch.

_____ **23.** Elements for forming a knot are:
   A. bight, loop, and round turn.
   B. loop, bend, and crown.
   C. round turn, standing, and running.
   D. standing, bight, and hitch.

_____ **24.** The end of the rope that is used to tie a knot is called the:
   A. running end.
   B. standing part.
   C. safety end.
   D. working end.

_____ **25.** A pike pole should:
   A. be hoisted point down.
   B. be hoisted sideways.
   C. be hoisted point up.
   D. not be hoisted due to the risk of injury to fire fighters on the ground.

_____ **26.** A tag/guide line should be used when:
   A. an overhang exists on which the item being hoisted is likely to get caught.
   B. there is a strong wind that may cause the item being hoisted to blow out of control.
   C. the item being hoisted may rub against the side of the structure causing damage.
   D. All of the above.

_____ **27.** Which of the following **best** describes a bight?
   A. The part of the rope used for the work
   B. The part of the rope used for hoisting or pulling
   C. Bending one end of the rope back upon itself, while keeping the two sides of the rope parallel
   D. Forming a loop around to a plane parallel with the other side

_____ **28.** The drawing below depicts one of the three elements of a knot or hitch. It is known as a:

   A. bight.

   B. loop.

   C. round turn.

   D. becket/sheet bend.

_____ **29.** Which one of the following statements is <u>**incorrect**</u> with regard to natural fiber ropes?

   A. When compared to synthetic materials, they have a very low strength-to-weight ratio.

   B. They work best as life safety lines.

   C. They have a low shock load absorption capability.

   D. They suffer from natural deterioration/degradation due to age.

_____ **30.** The drawing below depicts one of the three elements of a knot or hitch. It is known as a:

   A. bight.

   B. loop.

   C. round turn.

   D. clove hitch.

_____ **31.** The **<u>first</u> <u>step</u>** in tying the becket/sheet bend is to form a:
A. loop.
B. bight.
C. round turn.
D. half hitch.

_____ **32.** The stage phases of fire development are:
A. initial, growth, developed, decline.
B. incipient/ignition, growth, fully developed, decay.
C. origin, intermediate, growth, decline.
D. spontaneous, incipient, growth, decay.

_____ **33.** All fire department radio operations must follow rules of the:
A. NFPA.
B. DOT.
C. FCC.
D. CFR.

_____ **34.** All of the following are considered valuable characteristics or traits for a person who receives emergency calls **<u>except</u>**:
A. the ability to perform multiple tasks.
B. an inability to retain composure.
C. the ability to remember details and recall information easily.
D. the ability to exercise voice control.

_____ **35.** Determining exactly which units to send to a call depends on the location of the incident and the:
A. time since last response (TSLR) of the nearest units.
B. time lapse between call receipt and dispatch.
C. classification of the incident.
D. identity of the caller.

_____ **36.** Computer-aided dispatch (CAD) can be defined as a/an:
A. computer-based automated system that assists the telecommunicator in assessing dispatch information and recommends responses.
B. organized collection of similar facts.
C. system typically used by operations chief officers in the fire service.
D. emergency alerting devices primarily used by volunteer department personnel to receive reports of emergency incidents.

_____ **37.** When receiving reports of emergencies by telephone, the individual should always speak:
A. rapidly with low volume.
B. softly with some hesitation.
C. clearly, slowly, and with good volume.
D. clearly, rapidly, and with good volume.

_____ **38.** What is the largest difference between Basic 911 and Enhanced 911?
  A. Enhanced systems have the capability to provide the caller's telephone number and address.
  B. Enhanced systems are used only in rural areas.
  C. Basic systems are more reliable than enhanced.
  D. Basic systems have the capability to provide the caller's telephone number and address.

_____ **39.** Complete and accurate records should be maintained at communication centers for:
  A. all responses.
  B. only emergency responses.
  C. only responses that may be criminally related.
  D. areas of the district that generate high call volume.

_____ **40.** The **most** **important** piece of information that the caller gives is:
  A. from where he/she is calling.
  B. the caller's telephone number.
  C. the location of the emergency.
  D. the nature of the emergency.

_____ **41.** Information needed to determine the scope of emergency includes:
  A. incident location.
  B. incident/situation.
  C. time the incident occurred.
  D. All of the above.

_____ **42.** Which of the following **is** **not** a proper radio procedure for fire personnel?
  A. Transmit when the airwaves are clear.
  B. Hold the radio/microphone one to two inches from the mouth.
  C. Speak as you key the microphone to save time.
  D. Think about what is going to be said prior to transmitting.

_____ **43.** During a fire, you hear another team call "Mayday." You should:
  A. report on the radio to your supervisor advising of your location.
  B. stay off the radio and listen for instructions.
  C. rush into the building to find the crew calling for help.
  D. activate your emergency button on your radio.

_____ **44.** Before transmitting any information over the fire department radio, a fire fighter should:
  A. press the key two or three times to signal the intent to transmit.
  B. key the microphone and then clear his/her throat to be sure his/her voice will be clear.
  C. turn up the volume control if his/her voice is naturally soft.
  D. listen to be sure the channel is **not** being used.

_____ **45.** The coupling on the high pressure hose that is attached to an SCBA cylinder should be:
A. of the reverse thread type.
B. hand tight.
C. secured with an adjustable wrench.
D. treated with teflon tape.

_____ **46.** The bypass valve on a self-contained breathing apparatus is used:
A. during hazardous materials incidents.
B. in emergency situations involving a malfunctioning regulator.
C. to clear the mask of unwanted condensation.
D. to cool the face piece when high heat is encountered.

_____ **47.** Which of the following is considered to be a hazardous atmosphere encountered during fires?
A. Oxygen deficiency
B. Elevated temperatures
C. Smoke
D. All of the above.

_____ **48.** When tightening the straps on a SCBA face piece, the _____ straps should always be tightened first.
A. lower
B. temple
C. upper
D. harness

_____ **49.** Four hazardous atmospheres that fire fighters are likely to encounter at a fire are:
A. super-heated air, toxic gases, oxygen deficiency, and smoke.
B. toxic gases, hyperventilation, oxygen deficiency, heat.
C. heat, light, smoke, and chemical chain reaction.
D. carbon monoxide, ammonia, water, and hydrogen sulfide.

_____ **50.** Rescue from sewers, grain silos, and similarly confined spaces requires the use of self-contained breathing apparatus due to the danger of:
A. toxic gases.
B. oxygen deficiency.
C. ambient temperature.
D. Both A and B are correct.

_____ **51.** PASS devices are designed to assist rescuers attempting to:
A. move through traffic while responding to an incident.
B. locate trapped fire fighters.
C. eject smoke from a building.
D. roll hose faster than by hand.

_____ **52.** Which type of breathing apparatus recycles the user's exhaled breath after removing carbon dioxide and adding supplemental oxygen?
  A. Open-circuit
  B. Closed-circuit
  C. SAR
  D. Positive pressure

_____ **53.** What are the two general types of self-contained breathing apparatus?
  A. Demand and pressure-demand
  B. Open-circuit and closed-circuit
  C. OSHA approved and NIOSH approved
  D. Compressed air and liquid oxygen

_____ **54.** Which of the following **is** **not** a component of an open-circuit breathing apparatus?
  A. Regulator
  B. Face piece
  C. Low-pressure alarm
  D. Oxygen cylinder

_____ **55.** Which of the following **is** **true** regarding SCBA?
  A. Tighten the top straps of the SCBA face piece first.
  B. All SCBA face pieces must be fit-tested.
  C. Hoods should be worn under the SCBA straps and face piece.
  D. Positive pressure should be tested by breaking the regulator seal.

_____ **56.** One requirement of structural fire fighting gloves is that they must:
  A. provide a positive electrical ground.
  B. have three points of flexion.
  C. meet the applicable NFPA standard.
  D. be able to be twisted and wrung dry.

_____ **57.** Failure to wear your PPE can lead to:
  A. negligence.
  B. injury.
  C. fines.
  D. increased protection.

_____ **58.** Which of the following **is** **not** one of the advantages to wearing gloves that meet NFPA standards for structural firefighting?
  A. Thermal protection
  B. Protection from puncture
  C. Protection from scrapes and cuts
  D. Reduced dexterity

_____ **59.** The incident action plan is based on:
  A. measured uncertainties.
  B. information gathered during sizeup.
  C. projected initial fire history.
  D. principles of thermodynamics.

_____ **60.** A flat-head ax is more suitable for _____, while a pick-head ax is more adaptable to a variety of firefighting functions.
  A. striking
  B. prying
  C. heavy work
  D. chopping

_____ **61.** The K-tool is used in conjunction with a:
  A. Halligan tool.
  B. battering ram.
  C. aerial device.
  D. pike pole.

_____ **62.** Fire fighters can expect to find ledge doors in:
  A. barns and warehouses.
  B. single-family residential buildings.
  C. churches and temples.
  D. commercial occupancies.

_____ **63.** One way to force a lock is to physically pull the _____ out of the door using an A-tool or a K-tool.
  A. keyhole
  B. cylinder
  C. hasp
  D. strike plate

_____ **64.** Of the choices listed below, the **least desirable** point to force entry into a structure is:
  A. a check rail window.
  B. a swinging door.
  C. plate glass.
  D. a revolving door.

_____ **65.** Panel, slab, and ledge are all types of _____ doors.
  A. wood swinging
  B. metal swinging
  C. overhead rolling
  D. revolving

_____ **66.** With a pry bar/rambar, the bar acts as a _____ to multiply the amount of force the person could otherwise apply.
  A. mechanical
  B. gear
  C. spring
  D. lever

_____ **67.** Once a fire fighter has broken a window for purposes of entry, the next action should be to:
  A. call for a charged line.
  B. carefully climb through the window.
  C. open a window on the windward side of the building.
  D. clear the entire window area of glass.

_____ **68.** Because of the high risks associated with search and rescue:
  A. all floors must be searched simultaneously.
  B. it is done by teams of fire fighters.
  C. one fire fighter should accomplish the task.
  D. it is **not** recommended in large structures.

_____ **69.** Safety requires that fire fighters using self-contained breathing apparatus must work:
  A. alone.
  B. in contained areas.
  C. in pairs.
  D. with a lifeline.

_____ **70.** Before fire fighters enter a burning building to perform rescue work, they must first consider:
  A. manpower on the scene.
  B. weather conditions.
  C. damaging evidence of forced entry.
  D. their own safety.

_____ **71.** A personnel accountability system tracks which fire fighters are assigned to each vehicle and:
  A. when each fire fighter arrives.
  B. each crew's assignment.
  C. intra-crew communications.
  D. post-dispatch caller information.

_____ **72.** What is one basic method for staying oriented inside a low-visibility environment?
  A. Staying in contact with the hose line
  B. Counting doorways
  C. Sounding the floor periodically with an ax
  D. Counting paces or crawl movements

_____ **73.** A _____ System helps the Incident Commander know who is on the fireground and where fire fighters are located.
  A. Personnel Accountability
  B. Personal Alert
  C. Personnel Attendance
  D. P.A.S.S.

_____ **74.** One purpose of a Personnel Accountability System is to:
   A. keep track of fire fighters' salaries.
   B. identify trapped or injured fire fighters.
   C. create a job complaint form.
   D. help the Fire Chief control sick leave abuse.

_____ **75.** Which of the following <u>violates</u> a rule of personal safety?
   A. Always work in pairs or teams.
   B. Completely search one room before moving on.
   C. Remain standing even when you cannot see your feet.
   D. Before entering the building, locate more than one means of egress.

_____ **76.** In the fire service, the basic definition of the word "rescue" is:
   A. removing a victim from a hazardous situation to safety.
   B. stabilizing a victim before transporting.
   C. performing cardiopulmonary resuscitation(CPR) on a victim.
   D. All of the above.

_____ **77.** When lifting an object during a search, fire fighters should <u>**always**</u>:
   A. use their back to lift.
   B. use their legs to lift, not their back.
   C. try to twist and reach at the same time.
   D. lift with their arms and back.

_____ **78.** The _____ carry can be used on a conscious patient.
   A. fire fighter's
   B. leg
   C. seat
   D. two-person

_____ **79.** When being dragged, a patient should always be dragged:
   A. feet first.
   B. head first.
   C. sideways.
   D. after being completely immobilized.

_____ **80.** Which of the following statements is <u>**incorrect**</u>?
   A. Use care when dragging a fire fighter wearing an SCBA to prevent breaking the seal.
   B. All types of rescue drags provide good spinal immobilization.
   C. Drags are intended to be used where greater harm will come to the patient if he is not immediately moved.
   D. Rescue drags do <u>**not**</u> provide effective spinal immobilization.

_____ **81.** The area that is the <u>**first**</u> search priority is the:
   A. area immediately around the fire.
   B. exterior of the structure.
   C. top floor.
   D. area directly inside the main egress.

_____ **82.** A hose roller is a:
   A. rope or strap used to hoist hose.
   B. device by which a hose can be rolled.
   C. device that is fastened to the hose coupling in preparation for fastening.
   D. curved metal frame that fits over the edge of a roof or windowsill.

_____ **83.** Solid-stream hand lines are designed to be operated at a nozzle pressure of _____ psi.
   A. 50
   B. 75
   C. 90
   D. 100

_____ **84.** During an attack on a confined interior building fire, water **should not** be applied for too long or the _____ will be upset, causing it to drop rapidly to the floor.
   A. thermal layer
   B. vapor layer
   C. flammable limit
   D. combustion layer

_____ **85.** In reference to a fire stream, friction loss is defined as:
   A. the loss of the stream velocity after it exits the nozzle.
   B. that part of the pressure that is used to overcome friction in the hose.
   C. the amount of pressure needed to overcome friction caused by water turbulence.
   D. the friction created by dragging the hose along the ground.

_____ **86.** Devices through which water flows, used in conjunction with fire hose, such as a gate valve, are known as:
   A. apparatus.
   B. appliances.
   C. tools.
   D. flow controls.

_____ **87.** The standpipe connection is usually located in the _____ of a multistory building.
   A. equipment room
   B. building lobby
   C. elevator shaft (bottom)
   D. stairwell

_____ **88.** Which of the following statements regarding the accordion load is **incorrect**?
   A. It is easy to load and unload.
   B. It can be used as an additional supply line.
   C. It is ideal for making shoulder loads.
   D. It is the preferred way to load large diameter hose (LDH).

_____ **89.** The first due engine company's purpose on the fire scene is:
  A. ventilation.
  B. salvage and overhaul.
  C. to advance hose lines.
  D. forcible entry.

_____ **90.** Which of the following statements regarding advancing a hose line up a ladder is <u>incorrect</u>?
  A. The safest and best manner is to advance a charged hose line up the ladder.
  B. The safest and best manner is to advance an uncharged hose line up the ladder.
  C. Use two fire fighters who coordinate and maintain their distance on the ladder and the hose.
  D. The nozzle person, with both hands free for climbing, begins climbing the ladder to point about 20 feet up and stops.

_____ **91.** Cellar nozzles are designed to be used to fight localized fires in basements where fire fighters cannot make a direct attack. It operates by:
  A. forming a water curtain to protect exposures.
  B. directing the stream in one direction.
  C. piercing block walls or aircraft skin.
  D. rotating in a circular spray pattern.

_____ **92.** The purpose of a hose adaptor is to connect two hose sections of:
  A. equal diameter, but differing threads.
  B. unequal diameter but similar threads.
  C. equal diameter and similar threads.
  D. unequal diameter and differing threads.

_____ **93.** The type of appliance used to attach hose lines of differing diameters is a/an:
  A. adaptor.
  B. engager.
  C. distributor.
  D. reducer.

_____ **94.** Proper ventilation reduces danger of asphyxiation, enhances visibility, and removes:
  A. overhaul concerns.
  B. excess moisture.
  C. salvage concerns.
  D. heat.

_____ **95.** One <u>disadvantage</u> of a positive-pressure fan is that they:
  A. can spread the fire if used improperly.
  B. require the products of combustion to pass through them.
  C. block up an otherwise usable doorway.
  D. require a team of fire fighters several minutes to hang, set up, and seal.

_____ **96.** The <u>primary</u> function of smoke ejectors or exhaust fans is:
   A. localizing the fire.
   B. removing heat and smoke.
   C. providing fresh air for attack crews.
   D. removing lighter-than-air gases.

_____ **97.** Forced/mechanical ventilation is accomplished by blowers, fans, or:
   A. removal of windows.
   B. vertical openings.
   C. fog streams.
   D. natural wind currents.

_____ **98.** When a window is opened for the purpose of ventilation:
   A. screens may be left in place.
   B. curtains or drapes may be left in place.
   C. screens must be removed.
   D. it signifies that search and rescue has been accomplished.

_____ **99.** Convection is:
   A. transfer of heat through space by infrared rays.
   B. transfer of heat through a solid medium.
   C. <u>not</u> considered a method of heat transfer.
   D. transfer of heat through liquids or gases by circulating currents.

_____**100.** Of the following tools used in ventilation operations,
   a _____ would be <u>best</u> for sounding the roof.
   A. nozzle
   B. pick-head ax
   C. power saw with extended chain bar
   D. truss finder

_____ **101.** When cutting through a roof, a fire fighter should attempt to:
   A. remove the ceiling joist in the ventilation hole.
   B. cut a large circular hole.
   C. make the opening square or rectangular.
   D. stand to the downwind side.

_____**102.** During salvage and overhaul operations, it is essential for fire fighters to:
   A. remove their coats once the fire has been extinguished.
   B. work without coat, gloves, and helmet.
   C. remove protective breathing apparatus.
   D. wear complete protective equipment, including SCBA.

_____**103.** A tool often used to open a ceiling is a:
   A. pick-head ax.
   B. pike pole.
   C. kelly tool.
   D. K-tool.

_____**104.** Hidden fires in concealed spaces can often be detected by:
   A. feeling with back of hand.
   B. strategic fan placement.
   C. tearing down the entire wall.
   D. use of salvage techniques.

_____**105.** Which of the following **is** **not** considered a valuable benefit of proper overhaul?
   A. Helps locate hidden fires
   B. Helps prevent rekindle
   C. Helps in the removal of water
   D. All of the above.

_____**106.** Fire fighters can assist the investigator in all the following ways **except** by:
   A. performing overhaul before the investigator arrives.
   B. reporting unusual fire behavior.
   C. waiting for the investigator to release an area for overhaul.
   D. reporting what witnesses to the fire have said about the fire starting.

_____**107.** A commonly used method for two fire fighters to deploy a large salvage cover is a/an:
   A. combination throw/toss.
   B. balloon throw/toss.
   C. accordion toss.
   D. horseshoe throw/toss.

_____**108.** Fire protection professionals agree that salvage work is:
   A. **not** practical considering present day staffing requirements.
   B. **not** a fire department responsibility.
   C. an effective means of promoting positive public relations.
   D. best accomplished by private companies who specialize in this kind of work.

_____**109.** To form the corners of the basin when constructing a catchall, the fire fighter should lay ends of the side rolls over at a _____ angle.
   A. 30°
   B. 60°
   C. 90°
   D. 180°

_____**110.** A carry-all is used for:
   A. removing fire victims.
   B. bringing additional fire equipment to the scene.
   C. catching water that is leaking into an area.
   D. carrying debris out of an area.

_____**111.** What feature is required on a fire department connection that allows a fire fighter to charge the first hose line prior to connecting the second?
   A. Gate valve
   B. Clapper valve
   C. Quarter-turn couplings
   D. Cap

_____ **112.** If individual sprinkler heads **cannot be** shut off, what should be done to stop the flow of water?
   A. Crimp the piping in the riser.
   B. Open the fire department connection outside.
   C. Close the sprinkler control valve.
   D. Activate more heads using a mechanical heat source.

_____ **113.** One limitation of backpack water vacuums is that they:
   A. cannot pull water up from the floor.
   B. cannot be worn with SCBA.
   C. can only be used once.
   D. are illegal in many places.

_____ **114.** What is the type of heat detector called which activates if the temperature of the surrounding air rises more than a set amount in a given period of time?
   A. Closed-chamber heat detector
   B. Rate-of-rise heat detector
   C. Open-chamber heat detector
   D. Variable-threshold heat detector

_____ **115.** After connecting the supply line to the hydrant, the fire fighter should _____ the hydrant.
   A. partially open
   B. fully open
   C. barely crack open
   D. never open

_____ **116.** When performing a forward lay, the pumper is driven from the:
   A. fire scene to water source.
   B. water source to fire scene.
   C. water source to fire scene to water source.
   D. discharge of one pumper to intake of another.

_____ **117.** The term reverse lay describes an apparatus that lays out a supply line:
   A. while the apparatus is moving in reverse.
   B. from the fire to the water source.
   C. from the water source to the fire.
   D. with the male coupling ending up at the fire scene.

_____ **118.** Soft sleeve intake hose is used for:
   A. transferring water from a hydrant to an apparatus.
   B. primarily drafting water from a static source.
   C. siphoning water from one portable tank to another.
   D. transferring water from pump to tank.

_____ **119.** An advantage of a forward lay is:
   A. that the pumper is located at the hydrant to boost water pressure.
   B. that the supply line is dropped off at the fire location.
   C. the ability to utilize poor or static water sources.
   D. that the pumper is located at the fire with access to additional hose lines.

_____120. The space provided for hose on fire apparatus is generally referred to as the:
   A. hose box.
   B. hose load.
   C. hose bed.
   D. hose lay.

_____121. Portable water tanks should be positioned in a location that allows easy access from:
   A. multiple directions.
   B. only one direction.
   C. the windward side.
   D. the leeward side.

_____122. The **most** **common** water distribution system is a _____ system.
   A. pumped
   B. combination pumped/gravity
   C. gravity
   D. tender shuttle

_____123. Which of the following statements regarding dead-end water mains is **incorrect**?
   A. Connecting to a dead-end main may not provide adequate water.
   B. Two units on the same dead-end main may rob water from each other.
   C. Dead-end mains provide the most volume in the system.
   D. Where possible, fire fighters should avoid using them.

_____124. Which of the following statements regarding wet-barrel and dry-barrel hydrants is **incorrect**?
   A. Dry-barrel hydrants are used in areas where freezing temperature could damage the hydrant.
   B. Wet-barrel hydrants have water in the barrel up to the valves of each outlet.
   C. Dry-barrel hydrants use a valve at the base to control water flow to all outlets.
   D. Wet-barrel hydrants are commonly used in the northern parts of the United States.

_____125. A blue circle with a letter designation in the center would indicate an extinguisher is rated for use on _____ fires.
   A. Class A
   B. Class B
   C. Class C
   D. Class D

_____126. Class B fires involve fuels such as:
   A. flammable liquids.
   B. energized electrical equipment.
   C. combustible metals.
   D. ordinary combustibles.

_____**127.** Energized electrical equipment and the surrounding area have to be protected with extinguishers that have a _____ rating.
  A. Class A
  B. Class B
  C. Class C
  D. Class D

_____**128.** Extinguishers suitable for Class A fires can be identified by a _____ containing the letter "A." If colored, it should be _____.
  A. circle, blue
  B. star, yellow
  C. triangle, green
  D. square, red

_____**129.** A stored-pressure water extinguisher should be chosen to attack a _____ fire.
  A. Class A
  B. Class B
  C. Class C
  D. Class D

_____**130.** Dry powder extinguishers are rated for use on _____ fires.
  A. Class A
  B. Class B
  C. Class C
  D. Class D

_____ **131.** Halon fire extinguishers are primarily designed for use on _____ fires.
  A. Class A, B, and C & O
  B. Class B and C
  C. Class A and B
  D. Class C only

_____**132.** Class C fires can be extinguished using:
  A. multipurpose dry chemicals.
  B. water.
  C. water-based agents or foam.
  D. compressed air aspirating foam.

_____**133.** What factors should be considered when selecting a fire extinguisher?
  A. The location of the extinguisher
  B. The fuel
  C. The weight of the extinguisher
  D. The response time of the fire department

_____**134.** Clean agents are:
   A. materials specially formulated to clean fire extinguishers.
   B. gases that do **not** conduct electricity or leave a residue and are nonvolatile.
   C. agents that have been banned for causing ozone depletion.
   D. materials used to clean grease and oil off the floor.

_____**135.** Pump tank extinguishers are used to apply:
   A. water.
   B. wet chemical.
   C. dry chemical.
   D. carbon dioxide.

_____**136.** Class K fires involve:
   A. combustible metals.
   B. wood and paper.
   C. combustible cooking oils.
   D. energized equipment.

_____**137.** When faced with an electrical emergency the fire fighter shall try to:
   A. obtain a pair of lineman gloves.
   B. have equipment de-energized.
   C. use a dry rope to pull victim from contact with an energized conductor.
   D. wear rubber boots when approaching the emergency scene.

_____**138.** Fire fighters should treat all downed wires as:
   A. energized.
   B. safe if in contact with the ground.
   C. only dangerous if nearby homes have power.
   D. safe if **not** arcing.

_____**139.** The **safest** way for a fire fighter to disconnect electrical service to a building is to:
   A. remove the electrical meter.
   B. cut the service entrance wires.
   C. shut off the main breakers/switch at the service panel.
   D. pull the main breaker at the power pole.

_____**140.** Metal ladders that have been exposed to excessive heat should be:
   A. thoroughly washed before being placed back into service.
   B. visually inspected for metal fatigue.
   C. constructed well enough to withstand damage.
   D. placed out of service until tested.

_____**141.** What is the **usual** method of choice for drying ropes?
   A. Use of mechanical drying device
   B. In-the-bag drying
   C. Air drying
   D. Stretching rope very taut

_____**142.** When drying synthetic rope, keep it:
   A. in a warm, dry, sunlit place.
   B. in a cool, moist, sunlit place.
   C. out of sunlight.
   D. in a damp, dark place.

_____**143.** When inspecting a kernmantle rope, what finding should alert fire fighters to possible damage in the rope?
   A. Color change of the kern
   B. Apparent intactness of the mantle
   C. Depressions of the kern
   D. Color change of the stress threading in the mantle

_____**144.** A fire department's comprehensive SCBA program should include:
   A. inspecting, disinfecting, maintaining, and storing.
   B. visual inspection of the harness and frame only.
   C. an annual maintenance as a minimum.
   D. using, recording, cleaning, and examining only.

_____**145.** Halyards and wire cables on extension ground ladders shall be replaced:
   A. annually.
   B. semiannually.
   C. by certified technicians.
   D. when they become frayed or kinked.

_____**146.** _____ should be applied to wooden tool handles to prevent roughness and warping.
   A. Paint
   B. Mild soap
   C. Boiled linseed oil
   D. Varnish

_____**147.** When hose has been exposed to small amounts of oil, it should be washed with:
   A. clear water.
   B. cold water and scrub brush.
   C. mild soap or detergent.
   D. solvent solution.

_____**148.** The main advantage of the donut roll is:
   A. the female end is exposed and the male end is protected in the center of the roll.
   B. both ends are available on the outside of the roll.
   C. the hose can be rolled into a twin roll and secured by a portion of the hose itself.
   D. the couplings are connected to keep them together and protect the threads.

_____**149.** Hose is unloaded at the scene of a fire and the pumper proceeds to the water source. This is an example of a _____ lay.
    A. combination
    B. forward
    C. reverse
    D. direct

_____**150.** Which of the following **is not** a method for preventing mechanical damage to fire hose?
    A. Avoid closing the nozzle abruptly.
    B. Remove wet hose from apparatus and replace with dry hose.
    C. Prevent vehicles from driving over fire hose.
    D. Avoid laying hose over rough, sharp edges or corners.

——————— **Helpful Hint** ———————

Try to determine why you selected the wrong answer. Usually something influenced your selection. Focus on the difference between your wrong answer and the correct answer. Carefully read and study the entire paragraph containing the correct answer. Highlight the answer just as you did for Examination I-1.

Did you score higher than 80 percent on Examination I-2? Circle Yes or No in ink.

Now that you have finished the feedback step for Examination I-2, it is time to repeat the process by taking another comprehensive examination for NFPA 1001 Standard.

Examination I-3, Confirming What You Mastered

During Examination I-3, progress will be made in reinforcing what you have learned and improving your examination-taking skills. This examination contains approximately 60 percent of the examination items you have already answered and several new examination items. Follow the steps carefully to realize the best return on effort.

Step 1—Take Examination I-3. When you have completed Examination I-3, go to Appendix A and compare your answers with the correct answers.

Step 2—Score Examination I-3. How many examination items did you miss? Write the number of missed examination items in the blank in ink _____. Enter the number of examination items you guessed in this blank _____. Enter these numbers in the designated locations on your Personal Progress Plotter.

Step 3—During the feedback step, research the correct answer using the Appendix A information for Examination I-3. Highlight the correct answer during your research of the reference materials. Read the entire paragraph containing the correct answer.

Examination I-3

Directions

Remove Examination I-3 from the manual. First, take a careful look at the examination. There should be 150 examination items. Notice that a blank line precedes each examination item number. This line is provided for you to enter the answer to the examination item. Write the answer in ink. Remember the rule about not changing your answers. Our research shows that changed answers are most often changed to an incorrect answer, and, more often than not, the answer that is chosen first is correct.

If you guess an answer, place an "X" or a check mark by your answer. This step is vitally important to gain and master knowledge. We will explain how we treat the "guessed" items later in SAEP.

Take the examination. Once you complete it, go to Appendix A and score your examination. Once the examination is scored, carefully follow the directions for feedback of the missed and guessed examination items.

_____ **1.** What is the leading cause of fire fighter injuries?
  A. Exposure to fire products
  B. Being struck by objects
  C. Overexertion and strain
  D. Exposure to chemicals

_____ **2.** What is developed to provide definite guidelines for present and future actions?
  A. Analyses
  B. Policies
  C. Comparisons
  D. Post-incident plans

_____ **3.** About how much of a kernmantle rope's strength is provided by the mantle?
  A. 10 to 15 percent
  B. 25 to 30 percent
  C. 45 to 50 percent
  D. 60 to 75 percent

_____ **4.** Rope should be protected from prolonged exposure to sunlight because:
  A. exposure to ultraviolet radiation can damage rope.
  B. the inside of a rope heats up to damaging temperatures.
  C. the heat causes delamination of the sheath of ropes.
  D. the rope can easily become hot enough to the touch to cause burns.

_____ **5.** Another term for a safety knot is a/an:
  A. rescue knot.
  B. finishing knot.
  C. overhand knot.
  D. double hitch.

_____ **6.** Who should always plan ahead so that there is a fresh or rested crew ready to rotate with a crew that needs rehabilitation?
 A. The Staging Officer
 B. The Company Officer
 C. The Incident Commander
 D. The Logistics Section Chief

_____ **7.** How soon after a traumatic call should Critical Incident Stress Debriefing be held?
 A. As soon as possible
 B. After returning to the station, showering, and calming down
 C. Some time during the next shift
 D. Not less than three days later, preferably in the evening

_____ **8.** The **primary** use of the _____ is to attach a rope to a round object such as a pike pole.
 A. clove hitch
 B. bowline
 C. becket/sheet bend
 D. rescue knot

_____ **9.** The _____ is tied as illustrated in the drawing below.
 A. bowline
 B. becket/sheet bend
 C. square knot
 D. follow through

_____ **10.** The end of the rope that is used to tie a knot is called the:
 A. running end.
 B. standing part.
 C. safety end.
 D. working end.

_____ **11.** Which of the following **best** describes a bight?
    A. The part of the rope used for the work
    B. The part of the rope used for hoisting or pulling
    C. Bending one end of the rope back upon itself, while keeping the two sides of the rope parallel
    D. Forming a loop around to a plane parallel with the other side

_____ **12.** Which one of the following statements is **incorrect** with regard to natural fiber ropes?
    A. When compared to synthetic materials, they have a very low strength-to-weight ratio.
    B. They work best as life safety lines.
    C. They have a low shock load absorption capability.
    D. They suffer from natural deterioration/degradation due to age.

_____ **13.** The **first step** in tying the becket/sheet bend is to form a:
    A. loop.
    B. bight.
    C. round turn.
    D. half hitch.

_____ **14.** The strategy of the incident includes which of the following?
    A. Offensive or defensive mode
    B. Rescue or fire suppression tactics
    C. Establish or pass command
    D. Incident stabilization or property conservation

_____ **15.** The ratio of the mass of a given volume of liquid compared with the mass of an equal volume of water at the same temperature is:
    A. specific gravity.
    B. vapor density.
    C. flash point.
    D. surface to mass ratio.

_____ **16.** All fire department radio operations must follow rules of the:
    A. NFPA.
    B. DOT.
    C. FCC.
    D. CFR.

_____ **17.** Computer-aided dispatch (CAD) can be defined as a/an:
    A. computer-based automated system that assists the telecommunicator in assessing dispatch information and recommends responses.
    B. organized collection of similar facts.
    C. system typically used by operations chief officers in the fire service.
    D. emergency alerting devices primarily used by volunteer department personnel to receive reports of emergency incidents.

_____. **18.** Information needed to determine the scope of emergency includes:
   A. incident location.
   B. incident/situation.
   C. time the incident occurred.
   D. All of the above.

_____ **19.** What does the Automatic Number Identification feature of Enhanced 911 do?
   A. Provide the caller's identity.
   B. Lock the caller's phone open.
   C. Show the caller's phone number.
   D. Provide access to real-time language translation.

_____ **20.** The telecommunicator's <u>first</u> responsibility is to:
   A. obtain the information that is required to dispatch the appropriate units to the correct location.
   B. reassure the caller that help will be forthcoming shortly and that his/her problem can be dealt with.
   C. provide the caller with real-time, emergency instructions such as how to perform CPR.
   D. ascertain the exact type of event that the caller is reporting.

_____ **21.** The coupling on the high pressure hose that is attached to an SCBA cylinder should be:
   A. of the reverse thread type.
   B. hand tight.
   C. secured with an adjustable wrench.
   D. treated with teflon tape.

_____ **22.** When using positive pressure SCBA, a poor seal between the face piece and the fire fighter's face is:
   A. <u>not</u> dangerous, because the positive pressure will keep toxic gases out of the face piece.
   B. dangerous, because it is depleting the air supply more quickly.
   C. <u>not</u> possible, because the positive pressure will seal the face piece to the face.
   D. the main cause of deaths on the fireground.

_____ **23.** As the oxygen supply in any given area falls below _____ percent, unconsciousness can occur.
   A. 21
   B. 17
   C. 12
   D. 9

_____ **24.** The <u>first</u> noticeable signs of oxygen deficiency are:
   A. profuse sweating and ringing in the ears.
   B. dizziness, impaired vision, and giddiness.
   C. increased respiratory rate and impaired muscular coordination.
   D. headache and rapid fatigue.

_____ **25.** Rescue from sewers, grain silos, and similarly confined spaces requires the use of self-contained breathing apparatus due to the danger of:
  A. toxic gases.
  B. oxygen deficiency.
  C. ambient temperature.
  D. Both A and B are correct.

_____ **26.** A fire fighter is susceptible to poisoning or irritation from carbon monoxide through:
  A. ingestion.
  B. absorption.
  C. inhalation.
  D. injection.

_____ **27.** Which of the following **is** **not** a limitation affecting a fire fighter's ability to use SCBA effectively?
  A. Physical
  B. Medical
  C. Mental
  D. Oxygen cylinder size

_____ **28.** When filling an SCBA cylinder, the cylinder must be:
  A. placed in a fragmentation containment or other shielded device.
  B. placed in water.
  C. filled in the open to allow for checking of signs of weakness in the cylinder.
  D. wrapped in a blanket or towel.

_____ **29.** Inhaled toxic gases can directly cause:
  A. disease of the lung tissue.
  B. muscle cramps in the lower extremities.
  C. blurred vision, leading to blindness.
  D. amnesia.

_____ **30.** Atmospheres are classified as oxygen deficient when they fall below _____ percent oxygen.
  A. 25
  B. 19.5
  C. 16
  D. 13.5

_____ **31.** Recharging air cylinders can be done from a bank of three or more large air cylinders. This type system is called a/an _____ system.
  A. combination
  B. cascade
  C. multiple
  D. in line

_____ **32.** The low alarm of the SCBA will sound when _____ percent of the cylinder is remaining.
  A. 15
  B. 20
  C. 25
  D. 30

_____ **33.** There are two methods that can be used to don SCBA that is stored in a case:
  A. over-the-head and coat method.
  B. compartment and coat method.
  C. over-the-head and compartment method.
  D. jacket and backup mount.

_____ **34.** Which of the following <u>is</u> <u>true</u> regarding SCBA?
  A. Tighten the top straps of the SCBA face piece first.
  B. All SCBA face pieces must be fit-tested.
  C. Hoods should be worn under the SCBA straps and face piece.
  D. Positive pressure should be tested by breaking the regulator seal.

_____ **35.** The purpose for the use of reflective trim on PPE is to:
  A. increase the visibility of the wearer to others.
  B. provide protection for material under the trim.
  C. allow the wearer to blend in with the surroundings.
  D. be more stylish than the plain PPE.

_____ **36.** Which of the following <u>is</u> <u>not</u> one of the advantages to wearing gloves that meet NFPA standards for structural firefighting?
  A. Thermal protection
  B. Protection from puncture
  C. Protection from scrapes and cuts
  D. Reduced dexterity

_____ **37.** Which of the following <u>is</u> <u>true</u> regarding personal protective clothing?
  A. It must comply with NFPA 1900 standard.
  B. It requires fire fighters to wear the most appropriate PPE for the hazard they face.
  C. It requires PASS activation whenever bunker gear is worn.
  D. It prevents a fire fighter from burns in situations such as flashover.

_____ **38.** What type of construction, found on some older wooden buildings, provides a path for rapid fire extension?
  A. Side split
  B. Bungalow
  C. Slate roof
  D. Balloon-frame

_____ **39.** Tempered plate glass should be broken only as a last resort. It is recommended that it be shattered by striking:
   A. in the center with a large, blunt object.
   B. at the lowest corner with the pick end of a fire ax.
   C. with a flat-headed ax with hands above the head.
   D. with a flat-headed ax at the highest corner.

_____ **40.** In what type of occupancy will fire fighters usually find sliding doors?
   A. Barns or warehouses
   B. Commercial
   C. Residential
   D. Institutional

_____ **41.** Fire fighters can expect to find ledge doors in:
   A. barns and warehouses.
   B. single-family residential buildings.
   C. churches and temples.
   D. commercial occupancies.

_____ **42.** A battering ram is intended to be used by two or four fire fighters. Which of the following describes its recommended use?
   A. It is held horizontally by all involved fire fighters, who run toward the object to be battered.
   B. If four fire fighters are involved, one guides the tip of the ram while two swing it in the direction of the fourth person.
   C. It is held horizontally by pairs of operators who swing it repeatedly against the object.
   D. It is held against the object to be battered by two fire fighters, while the third person pounds on the end with a sledge hammer.

_____ **43.** With a pry bar/rambar, the bar acts as a _____ to multiply the amount of force the person could otherwise apply.
   A. mechanical
   B. gear
   C. spring
   D. lever

_____ **44.** What is one example of a hand-powered hydraulic spreader tool?
   A. Spanner tool
   B. Pompier tool
   C. Rabbet tool
   D. Pilot tool

_____ **45.** A cutting torch has a flame temperature of **approximately**:
   A. 10,000°F.
   B. 15,000°F.
   C. 12,000°F.
   D. 5,000°F.

_____ **46.** Which of the following statements regarding the "K-tool" is <u>incorrect</u>?
    A. The front of the tool is shaped like the letter "K" and slides over the lock cylinder.
    B. The K-tool is designed to pull out lock cylinders and expose the mechanism in order to open the lock with the key tools.
    C. The back of the tool is shaped like the letter "K" and slides over the lock cylinder.
    D. The front of the tool has a loop for the adz of the Halligan tool.

_____ **47.** The quickest way to force entry through a security roll-up door is to:
    A. cut the door with a torch or saw.
    B. apply a K-tool to the latch.
    C. use a J-tool on the roller.
    D. use the irons.

_____ **48.** How many people are needed to use a battering ram?
    A. Two to four
    B. One or two
    C. Four to six
    D. Six only

_____ **49.** From a forcible entry standpoint, which of the following <u>is not</u> one of the ways doors function?
    A. Revolving
    B. Swinging
    C. Overhead
    D. Lifting

_____ **50.** Which of the following <u>is not</u> one of the basic types of locks?
    A. Mortise
    B. Bored (cylindrical)
    C. Rim
    D. Tumbled

_____ **51.** An important benefit of using a Personnel Accountability System is:
    A. knowing who is on the fireground.
    B. knowing which fire fighter has seniority.
    C. knowing which company arrived on the scene first.
    D. keeping track of which fire fighters work on which shift.

_____ **52.** In the system of accountability, what should be requested at the occurrence of tactical benchmarks, such as going from an offensive to a defensive strategy?
    A. A rehabilitation assignment for those coming off the fight
    B. A system communications check
    C. A PAR
    D. A BARKS check

_____ **53.** OSHA regulations require that when fire fighters are inside a structure fire, at least _____ trained and equipped fire fighters must be standing by outside.
  A. 2
  B. 3
  C. 4
  D. 6

_____ **54.** A fire fighter who is lost in a structure and finds a fire hose should remember:
  A. male coupling indicates the direction to the exit.
  B. female coupling indicates the direction of exit.
  C. activation of bypass is the best process for escape breathing.
  D. to immediately issue a PAR and then try to attempt a rescue.

_____ **55.** The parts of an extension ladder that prevent the fly section from being extended too far are called:
  A. guides.
  B. locks.
  C. anchors.
  D. stops.

_____ **56.** A bangor ladder has attachments for added leverage that are called:
  A. staypoles.
  B. truss poles.
  C. guide poles.
  D. rails.

_____ **57.** Other than a ladder safety belt, a fire fighter can be safely secured to a ground ladder using:
  A. a rope.
  B. an arm lock.
  C. hose strap.
  D. a leg lock.

_____ **58.** An upper section or top section of an extension ladder is known as a:
  A. truss.
  B. bed.
  C. fly.
  D. main.

_____ **59.** If 40 feet of a 50-foot extension ladder is used, the butt of the ladder should be placed approximately _____ feet from the building.
  A. 8
  B. 10
  C. 12
  D. 15

_____ **60.** The proper distance the foot or butt of a ladder should stand out from a building is _____ of the working distance of the ladder from the base of the wall.
- **A.** one-half
- **B.** one-quarter
- **C.** one-third
- **D.** one-eighth

_____ **61.** The very top of a ladder is called the:
- **A.** fly.
- **B.** top plate.
- **C.** claw.
- **D.** tip.

_____ **62.** The reason that ground ladders must be cleaned periodically even if **not used** at a scene is that:
- **A.** their heat-resistant treatment attracts dust.
- **B.** they build up an ionic charge which attracts dirt out of the air.
- **C.** normal atmospheric moisture reacts with them depositing a thin film.
- **D.** they accumulate dirt from being on the apparatus.

_____ **63.** Which of the following statements concerning ground ladder placement is correct?
- **A.** If a ladder is to be used for a fire fighter to break a window for ventilation, it should be placed on the leeward side.
- **B.** If a ladder is to be used for a fire fighter to break a window for ventilation, its tip should be even with the bottom of the window.
- **C.** If a ladder is to be used for entry or rescue from a window, its tip is placed slightly below the sill.
- **D.** If the window opening is wide enough to permit the ladder tip to project into it, place the ladder so that five rungs extend above the sill.

_____ **64.** Which of the following **is not** one of the recommended steps in fighting a fire in an electrically powered vehicle?
- **A.** Cut into high voltage components.
- **B.** Secure and stabilize the vehicle.
- **C.** Identify vehicle type.
- **D.** Use standard tools and cut in areas for victim extrication.

_____ **65.** Which of the following are characteristics of a **primary** search?
- **A.** Rapid and systematic
- **B.** Slow and deliberate
- **C.** Accomplished only with a charged hose line
- **D.** Begun only after top-side ventilation is provided

_____ **66.** When conducting a <u>primary</u> search within a structure, a fire fighter should begin:
A. in the center of the room.
B. on a wall.
C. always start with right hand pattern.
D. under or behind furnishings.

_____ **67.** When lifting an object during a search, fire fighters should <u>always</u>:
A. use their back to lift.
B. use their legs to lift, not their back.
C. try to twist and reach at the same time.
D. lift with their arms and back.

_____ **68.** In a multi-story building with a standpipe system, a fire fighter should make the connection _____ the fire floor.
A. on the floor below
B. on the floor above
C. on
D. two levels above

_____ **69.** One advantage of a solid stream nozzle is:
A. for ladder-pipe applications.
B. that it creates more stream.
C. its effectiveness in subduing stubborn Class B fires.
D. it provides an extended reach.

_____ **70.** To achieve extinguishment, sufficient water must be applied to:
A. thoroughly saturate the burning fuel.
B. smother all flames.
C. absorb heat faster than it is being generated.
D. produce steam.

_____ **71.** One way to identify the physical structure of a male coupling is by the:
A. swivel protector.
B. lugs on the shank.
C. coating.
D. shank length.

_____ **72.** The <u>primary</u> purpose of a spanner wrench is for use in:
A. breaking glass.
B. shutting-off gas valves.
C. operating hydrant valves.
D. tightening/loosening hose couplings.

_____ **73.** A hose _____ is used to seal small cuts or breaks that may occur in fire hose or to connect mismatched or damaged couplings of the same size to stop leaking.
A. bridge
B. clamp
C. jacket
D. seal

_____ **74.** A fire fighter **without** a spanner wrench can usually break a tight coupling if the _____ method is used.
  A. knee-jerk
  B. stiff-arm
  C. arm-press
  D. knee-press

_____ **75.** The first due engine company's purpose on the fire scene is:
  A. ventilation.
  B. salvage and overhaul.
  C. to advance hose lines.
  D. forcible entry.

_____ **76.** When should the fire fighter attaching hose to the hydrant charge the supply line?
  A. As soon as possible
  B. Not until the driver/operator's signal is received
  C. After slowly opening the hydrant all the way
  D. After ten full minutes from the time of drop-off

_____ **77.** A fire stream is:
  A. the flaming material shot through the air by a flame thrower.
  B. is the main attack line on interior attack.
  C. water or another agent as it leaves the nozzle toward the target.
  D. the total number of streams needed for knockdown and extinguishment of a fire.

_____ **78.** Hose that is **usually** carried on a hose reel, is made of rubber, and flows only 30 to 50 gallons per minute is called:
  A. booster hose.
  B. dumpster hose.
  C. fool's hose.
  D. attic hose.

_____ **79.** The main difference between types I and II construction is that:
  A. the occupants in type I construction are the main hazard.
  B. type I is more prone to building collapse.
  C. roofs in type II construction are more stable.
  D. structural components in type II construction lack insulation and protection.

_____ **80.** The primary concern of ordinary construction is:
  A. the exterior building materials.
  B. fire and smoke spreading through concealed space.
  C. early wall collapse.
  D. combustible contents and building materials.

_____ **81.** When a window is opened for the purpose of ventilation:
  A. screens may be left in place.
  B. curtains or drapes may be left in place.
  C. screens must be removed.
  D. it signifies that search and rescue has been accomplished.

_____ **82.** Heat transfer that takes place in the form of electromagnetic waves is known as:
   A. convection.
   B. radiation.
   C. conduction.
   D. flame contact.

_____ **83.** Heat can travel throughout a burning building by one or more of the following methods:
   A. conduction, radiation, and convection.
   B. extension, conduction, and exposure.
   C. convection, extension, and expansion.
   D. conduction, radiation, and expansion.

_____ **84.** The principle which **most** **closely** describes how water extinguishes fire is:
   A. removal of fuel.
   B. reduction of temperature.
   C. exclusion of oxygen.
   D. inhibition of chain reaction.

_____ **85.** A fire in the presence of a higher-than-normal concentration of oxygen will:
   A. burn slower than normal.
   B. burn faster than normal.
   C. **not** be effected by the oxygen.
   D. **not** burn if oxygen is too rich.

_____ **86.** Which of the following **are** **not** products of combustion?
   A. Convection, conduction, radiation, and direct contact
   B. Fire gases, heat, smoke, and light
   C. Fire gases, water vapor, and carbon particles
   D. Carbon monoxide, carbon dioxide, sulphur dioxide, and hydrogen cyanide

_____ **87.** The term vapor density refers to the weight of a gas as compared to the weight of:
   A. water.
   B. air.
   C. carbon.
   D. nitrogen.

_____ **88.** During _____, conditions in the compartment change very rapidly.
   A. rollover
   B. growth
   C. flashover
   D. fully-developed

_____ **89.** Just prior to flashover, what are the conditions within the
burning compartment?
  A. Temperatures are rapidly increasing.
  B. Additional fuel packages are becoming involved.
  C. Fuel packages in the compartment are giving off combustible gases.
  D. All the above are occurring.

_____ **90.** Which of the following gases **is** **not** produced in fires?
  A. Carbon monoxide
  B. Hydrogen chloride
  C. Carbon dioxide
  D. Oxygen

_____ **91.** Which of the following **is** **not** a benefit of proper ventilation?
  A. More oxygen is fed to the fire.
  B. Heat is removed from the structure.
  C. Visibility is improved by removing smoke from the structure.
  D. It allows occupants of the structure more survival time.

_____ **92.** By venting an enclosure, the heat level is kept from becoming capable
of producing:
  A. flashover.
  B. backdraft.
  C. rollover.
  D. Both A and B are correct.

_____ **93.** The size of an outside fire in the fully developed stage is dependent on:
  A. oxygen.
  B. temperature.
  C. available fuel.
  D. barometric pressure.

_____ **94.** Thermal layering is:
  A. a column of heat rising from a source.
  B. a process in which the molecules of a liquid are liberated into the
    atmosphere at a rate greater than the rate at which the molecules return
    to a liquid.
  C. the layered configuration of heat with higher temperatures at the upper
    levels and cooler temperatures at the lower levels.
  D. decomposition or transformation of a compound caused by heat.

_____ **95.** A Class A fire is fueled by:
  A. electricity.
  B. ordinary combustible materials.
  C. liquids.
  D. metals.

_____ **96.** Positive-pressure ventilation is:
- **A.** pulling smoke out.
- **B.** blowing fresh air in.
- **C.** removing oxygen.
- **D.** increasing thermal layering.

_____ **97.** An important safety precaution that should be practiced when working on a roof is to:
- **A.** cut all guy wires to prevent tripping over them.
- **B.** provide a secondary means of escape.
- **C.** have more than two fire fighters on the roof at all times.
- **D.** tie oneself to the roof ladder.

_____ **98.** The <u>most</u> <u>common</u> type of building material in use today is:
- **A.** steel.
- **B.** wood.
- **C.** concrete.
- **D.** plastics.

_____ **99.** What is one warning signal of possible backdraft conditions?
- **A.** Glass is smoke-stained and blackened due to heavy carbon deposits from the smoke.
- **B.** Smoke is observed pouring out of a burned-through opening in the roof.
- **C.** The front door is unaccountably wide open.
- **D.** Upper windows are observed to be open or shattered.

_____ **100.** The basic shape of the rigid framework of the truss is a/an:
- **A.** square.
- **B.** rectangle.
- **C.** triangle.
- **D.** oval.

_____ **101.** What does vertical ventilation require that horizontal ventilation does not?
- **A.** Positive-pressure fans
- **B.** Openings in the roof or highest part of the building
- **C.** Negative-pressure fans
- **D.** Airtightness everywhere but the vent opening

_____ **102.** Searching for hidden fires is a <u>primary</u> function of:
- **A.** sizeup.
- **B.** salvage.
- **C.** fire investigation.
- **D.** overhaul.

_____ **103.** Hidden fires in concealed spaces can often be detected by:
- **A.** feeling with back of hand.
- **B.** strategic fan placement.
- **C.** tearing down the entire wall.
- **D.** use of salvage techniques.

_____**104.** Ideally, when should fire investigators begin their work at a fire scene?
    A. After overhaul is complete
    B. Before overhaul starts
    C. Any time other than when overhaul operations are occurring
    D. During salvage operations

_____**105.** During overhaul, what should be available to douse any hot spots or flare-ups that occur?
    A. A Class D fire extinguisher
    B. A pike hole and Halligan bar
    C. A thermal blanket
    D. A charged hose line

_____**106.** Which areas of a wall are usually opened first, during overhaul?
    A. The most heavily damaged areas
    B. The least damaged areas
    C. The areas closest to any openings in that wall
    D. The area closest to the baseboard

_____**107.** Methods and operating procedures that reduce fire, water, and smoke damage during and after fires are known as:
    A. overhaul.
    B. sizeup.
    C. salvage.
    D. a coordinated fire attack.

_____**108.** In salvage operations, floor runners:
    A. are fire fighters who carry debris from the building.
    B. are water chutes constructed of rolled-up salvage covers placed to catch and drain excess water.
    C. is a term used to describe the type of fire which progresses vertically.
    D. are constructed of a lightweight, durable material placed over the floor to protect it from damage.

_____**109.** **Directions:** Read the statements below. Then choose the appropriate answer from choices A–D listed below.
    1. Salvage work in the fire service consists of procedures that reduce fire, water, and smoke damage during and after fires.
    2. Overhaul activities consist of all activities that take place after the fire as been extinguished.
    3. Overhaul operations must be completed before salvage.

        A. All statements are true.
        B. The first and second statements are false; the third statement is true.
        C. The first statement is true; the second and third statements are false.
        D. The first statement is false; the second and third statements are true.

_____ **110.** In addition to controlling runoff water, a salvage cover may be used:
   A. to collect debris.
   B. to cover furniture.
   C. as a catchall.
   D. All of the above.

_____ **111.** A fire department connection is nothing more than a:
   A. siamese.
   B. 2-1/2 inch wye.
   C. water thief.
   D. four-way valve.

_____ **112.** Which one of the following statements is <u>incorrect</u>?
   A. A floor runner is a fire fighter who acts as a messenger.
   B. The perimeter of the salvage cover is ringed with grommets.
   C. A floor runner is used to cover the floor in a hallway or along a traffic area.
   D. Water vacuums come in two basic types, backpack and wheeled.

_____ **113.** The temperature rating of a sprinkler head color-coded red is _____ degrees F.
   A. 175 to 225
   B. 250 to 300
   C. 325 to 375
   D. 400 to 475

_____ **114.** What technique is needed to use a sprinkler wedge to stop an activated sprinkler?
   A. Insert one wedge into the deflector and hammer it into place with helmet.
   B. Insert one wedge into the orifice and tap it into place with palm of hand.
   C. Insert one wedge into the orifice and one into the deflector, then deform them to touch by hammering sharply.
   D. Insert one wedge from either side between the orifice and the deflector and push them together.

_____ **115.** One limitation of backpack water vacuums is that they:
   A. cannot pull water up from the floor.
   B. cannot be worn with SCBA.
   C. can only be used once.
   D. are illegal in many places.

_____ **116.** What is the type of heat detector called which activates if the temperature of the surrounding air rises more than a set amount in a given period of time?
   A. Closed-chamber heat detector
   B. Rate-of-rise heat detector
   C. Open-chamber heat detector
   D. Variable-threshold heat detector

_____ **117.** When performing a forward lay, the pumper is driven from the:
    A. fire scene to water source.
    B. water source to fire scene.
    C. water source to fire scene to water source.
    D. discharge of one pumper to intake of another.

_____ **118.** Which of the following statements regarding dead-end water mains is **incorrect**?
    A. Connecting to a dead-end main may **not** provide adequate water.
    B. Two units on the same dead-end main may rob water from each other.
    C. Dead-end mains provide the most volume in the system.
    D. Where possible, fire fighters should avoid using them.

_____ **119.** All portable extinguishers are classified according to their:
    A. size.
    B. freeze potential.
    C. intended use.
    D. conductivity.

_____ **120.** CO_2 and dry chemical extinguishers will extinguish both Class B and C fires. What advantage does CO_2 have over a dry chemical extinguisher?
    A. CO_2 is **not** a hazard in an enclosed area.
    B. CO_2 does **not** leave a residue or corrode electrical contacts.
    C. CO_2 will prevent reignition longer than dry chemical extinguisher.
    D. CO_2 is effective at a greater distance.

_____ **121.** Class B fires involve fuels such as:
    A. flammable liquids.
    B. energized electrical equipment.
    C. combustible metals.
    D. ordinary combustibles.

_____ **122.** Energized electrical equipment and the surrounding area have to be protected with extinguishers that have a _____ rating.
    A. Class A
    B. Class B
    C. Class C
    D. Class D

_____ **123.** Fires involving combustible metals such as magnesium, titanium, zirconium, sodium, and potassium, are _____ fires.
    A. Class A
    B. Class B
    C. Class C
    D. Class D

_____ **124.** A carbon dioxide (CO_2) extinguisher's means of discharge is:
    A. chemical reaction.
    B. stored liquefied compressed gas.
    C. cartridge activation.
    D. manual hand-pump.

_____**125.** The manufacture of all _____ extinguishers has
been discontinued.
A. CO_2
B. dry chemical
C. inverting
D. pressurized-water

_____**126.** Regular dry chemicals (sodium bicarbonate-based and potassium-based) work
on _____ fires.
A. only Class D
B. Class A
C. only Class B
D. Classes B and C

_____**127.** What extinguishing agents are being replaced because they have been banned
for destroying the earth's ozone layer?
A. Halons or halogenated hydrocarbons
B. Carbon dioxide
C. Aqueous Film Forming Foam (AFFF)
D. Dry chemicals

_____**128.** What is the relationship between dry powder and dry chemical
extinguishing agents?
A. They are two different names for the same substance with the
same application.
B. Dry chemical is one of the many types of dry powder.
C. Dry powder is one of the many types of dry chemical.
D. They are entirely different substances with entirely different applications.

_____**129.** The first step for using a fire extinguisher is to:
A. aim the nozzle.
B. squeeze the handle.
C. pull the pin.
D. sweep the base of the fire.

_____**130.** One advantage of portable fire extinguishers over hose lines is that
fire extinguishers:
A. pack more suppression punch.
B. have controllable rates of flow.
C. don't run out as quickly.
D. are quicker to deploy and use.

_____**131.** Dry chemical extinguishers can be used on Class C fires because dry chemicals:
A. are chemically similar to water.
B. do not conduct electricity.
C. never actually touch the fuel.
D. are themselves electrically static.

_____**132.** What is one chemical used as a dry chemical extinguishing agent?
    A. Ammonium nitrate
    B. Tri-nitro toluene
    C. Methyl isocyanate
    D. Potassium bicarbonate

_____**133.** Class K fires involve:
    A. combustible metals.
    B. wood and paper.
    C. combustible cooking oils.
    D. energized equipment.

_____**134.** When faced with an electrical emergency the fire fighter shall try to:
    A. obtain a pair of lineman gloves.
    B. have equipment de-energized.
    C. use a dry rope to pull victim from contact with an energized conductor.
    D. wear rubber boots when approaching the emergency scene.

_____**135.** Which of the following **is** **least** likely to be shut off or disconnected by the fire fighter assigned to checking the utilities at a single family residential fire?
    A. Telephone/cable televisions
    B. Gas supply
    C. Domestic water
    D. Electric

_____**136.** The three **most important** factors that affect wildland firefighting are:
    A. fuel, equipment, and location.
    B. topography, resources, and time of day.
    C. fuel, weather, and topography.
    D. staffing, resources, and apparatus.

_____**137.** Worn, damaged, and deteriorated parts of a SCBA must be replaced according to:
    A. past practice.
    B. NIOSH/OSHA Respiratory Protection Act.
    C. manufacturer's instructions.
    D. the wearer's recommendations.

_____**138.** Composite SCBA cylinders must be hydrostatically tested every:
    A. year.
    B. three years.
    C. six years.
    D. ten years.

_____**139.** A fire department's comprehensive SCBA program should include:
    A. inspecting, disinfecting, maintaining, and storing.
    B. visual inspection of the harness and frame only.
    C. an annual maintenance as a minimum.
    D. using, recording, cleaning, and examining only.

_____**140.** _____ should be applied to wooden tool handles to prevent roughness and warping.
   A. Paint
   B. Mild soap
   C. Boiled linseed oil
   D. Varnish

_____**141.** After salvage covers are _____, they must be examined for holes and tears before being placed on the apparatus.
   A. rinsed off
   B. thrown
   C. dried
   D. rolled

_____**142.** It is permissible to use paint on a fire department ladder:
   A. to mark the bed section on multisection ladders.
   B. when there is a possibility of dry rot.
   C. to mark the ladder ends for visibility.
   D. when salt water may be a problem.

_____**143.** Fire department ladders should be _____ after each use and on a monthly basis.
   A. tested
   B. lubricated
   C. inspected
   D. varnished

_____**144.** When hose has been exposed to small amounts of oil, it should be washed with:
   A. clear water.
   B. cold water and scrub brush.
   C. mild soap or detergent.
   D. solvent solution.

_____**145.** When loading fire hose, the _____ serves to change the direction of the hose and can also change the location of a coupling.
   A. horseshoe
   B. flat load
   C. reverse lay
   D. Dutchman

_____**146.** The main advantage of the donut roll is:
   A. the female end is exposed and the male end is protected in the center of the roll.
   B. both ends are available on the outside of the roll.
   C. the hose can be rolled into a twin roll and secured by a portion of the hose itself.
   D. the couplings are connected to keep them together and protect the threads.

_____**147.** Hose is unloaded at the scene of a fire and the pumper proceeds to the water source. This is an example of a _____ lay.
  A. combination
  B. forward
  C. reverse
  D. direct

_____**148.** Which of the following **is** **not** a method for preventing mechanical damage to fire hose?
  A. Avoid closing the nozzle abruptly.
  B. Remove wet hose from apparatus and replace with dry hose.
  C. Prevent vehicles from driving over fire hose.
  D. Avoid laying hose over rough, sharp edges or corners.

_____**149.** Which coupling goes on the inside of a straight hose roll?
  A. The female
  B. The male
  C. Either A or B is correct.
  D. It depends on the diameter.

_____**150.** Which of the following **is** **not** a common load for preconnected attack lines?
  A. Flat load
  B. Triple layer load
  C. Straight load
  D. Minuteman load

Did you score higher than 80 percent on Examination I-3? Circle Yes or No in ink.

Feedback Step

Now, what do we do with your "yes" and "no" answers given throughout the Phase I process? First, return to any response that has "no" circled. Go back to the highlighted answers for those examination items missed. Read and study the paragraph preceding the location of the answer as well as the paragraph following the paragraph where the answer is located. This will expand your knowledge base for the missed question, put it in a broader perspective, and improve associative learning. Remember, you are trying to develop mastery of the required knowledge. Scoring 80 percent on an examination is good, but it is not mastery performance. To be at the top of your group, you must score much higher than 80 percent on your training, promotion, or certification examination.

Carefully review the Summary of Key Rules for Taking an Examination and Summary of Helpful Hints on the next two pages. Do this review now and at least two additional times prior to taking your next examination.

Helpful Hint

Studying the correct answers for missed items is a critical step in achieving your desired return on effort! The focus of attention is broadened and new knowledge is often gained by expanding association and contextual learning. During PTS's research and field test, self-study during this step of SAEP resulted in gains of 17 points between the first examination administered and the third examination. A gain of 17 points can move you from the lower middle to the top of the list of persons taking a training, promotion, or certification examination. That is a competitive edge and a prime example of return on effort in action. Remember: Maximum effort = maximum results!

Summary of Key Rules for Taking an Examination

<u>Rule 1</u>—Examination preparation is not easy. Preparation is 95% perspiration and 5% inspiration.

<u>Rule 2</u>—Follow the steps very carefully. Do not try to reinvent or shortcut the system. It really works just as it was designed to!

<u>Rule 3</u>—Mark with an "X" any examination items for which you guessed the answer. For maximum return on effort, you should also research any answer that you guessed even if you guessed correctly. Find the correct answer, highlight it, and then read the entire paragraph that contains the answer. Be honest and mark all questions on which you guessed. Some examinations have a correction for guessing built into the scoring process. The correction for guessing can reduce your final examination score. If you are guessing, you are not mastering the material.

<u>Rule 4</u>—Read questions twice if you have any misunderstanding, especially if the question contains complex directions or activities.

<u>Rule 5</u>—If you want someone to perform effectively and efficiently on the job, the training and testing program must be aligned to achieve this result.

<u>Rule 6</u>—When preparing examination items for job-specific requirements, the writer must be a subject matter expert with current experience at the level that the technical information is applied.

<u>Rule 7</u>—Good luck = good preparation.

Summary of Helpful Hints

<u>Helpful Hint</u>—Most of the time your first impression is the best. More than 41% of changed answers during our SAEP field test were changed from a right answer to a wrong answer. Another 33% were changed from a wrong answer to another wrong answer. Only 26% of changed answers were changed from wrong to right. In fact, three participants did not make a perfect score of 100% because they changed one right answer to a wrong one! Think twice before you change your answer. The odds are not in your favor.

<u>Helpful Hint</u>—Researching correct answers is one of the most important activities in SAEP. Locate the correct answer for all missed examination items. Highlight the correct answer. Then read the entire paragraph containing the answer. This will put the answer in context for you and provide important learning by association.

<u>Helpful Hint</u>—Proceed through all missed examination items using the same technique. Reading the entire paragraph improves retention of the information and helps you develop an association with the material and learn the correct answers. This step may sound simple. A major finding during the development and field testing of SAEP was that you learn from your mistakes.

<u>Helpful Hint</u>—Follow each step carefully to realize the best return on effort. Would you consider investing your money in a venture without some chance of earning a return on that investment? Examination preparation is no different. You are investing time and expecting a significant return for that time. If, indeed, time is money, then you are investing money and are due a return on that investment. Doing things right and doing the right things in examination preparation will ensure the maximum return on effort.

<u>Helpful Hint</u>—Try to determine why you selected the wrong answer. Usually something influenced your selection. Focus on the difference between your wrong answer and the correct answer. Carefully read and study the entire paragraph containing the correct answer. Highlight the answer.

<u>Helpful Hint</u>—Studying the correct answers for missed items is a critical step in achieving your desired return on effort! The focus of attention is broadened and new knowledge is often gained by expanding association and contextual learning. During PTS's research and field test, self-study during this step of SAEP resulted in gains of 17 points between the first examination administered and the third examination. A gain of 17 points can move you from the lower middle to the top of the list of persons taking a training, promotion, or certification examination. That is a competitive edge and a prime example of return on effort in action. Remember: Maximum effort = maximum results!

Fire Fighter II

Examination II-1, Surveying Weaknesses

At this point in SAEP, you should have the process of self-directed learning using examinations fixed in your mind. Moving through Phase II is accomplished in the same way as for Phase I. Do not attempt to skip steps in the process—after all, you now understand how SAEP works. Skipping steps can lead to a weak examination preparation result. The examinations will be more difficult in Phase II because of the increased level of required knowledge and skills. You will find that the SAEP methods move you gradually from the simple to the complex.

Do not study prior to taking the examination. Examination II-1 is designed to identify your weakest areas in terms of NFPA 1001. Some steps in SAEP will require self-study of specific reference materials. Remove Examination II-1 from the book.

Mark all answers in ink to ensure that no corrections or changes are made later. Do not mark through answers or change answers in any way once you have selected the answer. Doing so indicates uncertainty regarding the answer. Mastery is not compatible with uncertainty.

Step 1—Take Examination II-1. When you have completed Examination II-1, go to Appendix B and compare your answers with the correct answers. Each answer identifies reference materials with the relevant page numbers. If you missed the answer to the examination item, you have a source for conducting your correct answer research.

Step 2—Score Examination II-1. How many examination items did you miss? Write the number of missed examination items in the blank in ink _____. Enter the number of examination items you guessed in this blank _____. Enter these numbers in the designated locations on your Personal Progress Plotter.

Step 3—Now the learning begins! Carefully research the page cited in the reference material for the correct answer. For instance, use Jones and Bartlett, NFPA, *Fundamentals of Fire Fighter Skills, Second Edition*, go to the page number provided and find the answer.

Following are some of the rules and hints repeated from Phase I.

--------- Rule 3 ---------

Mark with an "X" any examination items for which you guessed the answer. For maximum return on effort, you should also research any answer that you guessed, even if you guessed correctly. Find the correct answer, highlight it, and then read the entire paragraph that contains the answer. Be honest and mark all questions on which you guessed. Some examinations have a correction for guessing built into the scoring process. The correction for guessing can reduce your final examination score. If you are guessing, you are not mastering the material.

——————————— **Rule 4** ———————————

Read questions twice if you have any misunderstanding, especially if the question contains complex directions or activities.

——————————— **Helpful Hint** ———————————

Proceed through all missed examination items using the same technique. Reading the entire paragraph improves retention of the information and helps you develop an association with the material and learn the correct answers. This step may sound simple. A major finding during the development and field testing of SAEP was that you learn from your mistakes.

Examination II-1

Directions

Remove Examination II-1 from the manual. First, take a careful look at the examination. There should be 100 examination items. Notice that a blank line precedes each examination item number. This line is provided for you to enter the answer to the examination item. Write the answer in ink. Remember the rule about not changing your answers. Our research shows that changed answers are most often changed to an incorrect answer, and, more often than not, the answer that is chosen first is correct.

If you guess the answer to a question, place an "X" or a check mark by your answer. This step is vitally important as you gain and master knowledge. We will explain how we treat the "guessed" items later in SAEP.

Take the examination. Once you complete it, go to Appendix B and score your examination. Once the examination is scored, carefully follow the directions for feedback on the missed and guessed examination items.

_____ 1. When a Company Officer arrives first on a fire scene, the officer is in command until:
   A. the fire is declared under control.
   B. a chief officer arrives and may choose to assume command.
   C. the chief of the department arrives.
   D. arrival of the senior shift officer.

_____ 2. An employee becomes frustrated because he/she cannot comply with conflicting orders from different bosses. This situation was caused by a violation of:
   A. chain of command.
   B. division of labor.
   C. span of control.
   D. unity of command.

_____ 3. Policies are examples of standing plans designed to provide:
   A. staffing requirement guidelines.
   B. guidance for decision making.
   C. problem-solving.
   D. communications.

_____ 4. A procedure is a/an:
   A. guide to thinking.
   B. detailed guide to action.
   C. guide to decision making.
   D. interpretation.

_____ 5. Under normal conditions, only a/an _____ may order multiple alarms or additional resources for large-scale incidents.
   A. logistics officer
   B. planning officer
   C. safety operations officer
   D. incident commander

_____ **6.** The tracking of personnel working at an incident requires a system that is standardized for every incident to establish:
  A. accountability.
  B. chain of command.
  C. unity of command.
  D. span of control.

_____ **7.** The Incident Command System is **best** defined as:
  A. the overall plan developed and used to control an incident.
  B. an organized, systematic method for the command, control, and management of an emergency incident.
  C. the ability to start small and expand if an incident becomes more complex.
  D. one designated leader or officer to command an incident.

_____ **8.** In order for the Incident Command System to function properly, it must contain all of the following components **except**:
  A. common terminology.
  B. integrated communications.
  C. all personnel from a single agency.
  D. consolidated incident action plans.

_____ **9.** A uniform data collection system used by most departments to track incident information is known as the:
  A. National Fire Incident Reporting System.
  B. National Fire Incident Recording System.
  C. First National Incident Reporting System.
  D. First National Incident Response System.

_____ **10.** The two **most common** ways the incident commander orders fire fighters to evacuate a structure are to broadcast a radio message and:
  A. page all fire fighters to respond.
  B. implement an accountability system.
  C. contact dispatch to activate PASS device.
  D. sound an audible warning.

_____ **11.** Audible warning devices for emergency evacuation should be:
  A. broadcast several times.
  B. heard for at least 500 feet.
  C. used to announce the need for multiple alarms.
  D. used to give an "all clear" on scene.

_____ **12.** It is important that the communication center be kept advised of the actions taken at emergency scenes. Situation status/progress reports should include all of the following **except**:
  A. change in command location.
  B. exposures present.
  C. direction of fire spread.
  D. number of units in staging.

_____ **13.** Fire departments that operate radio equipment must hold radio licenses from the:
  A. Federal Central Communications.
  B. National Emergency Broadcasting.
  C. Federal Communications Commission.
  D. National Radio Communications.

_____ **14.** In fire departments that have access to multiple radio channels, emergency operations should be:
  A. on multichannels also.
  B. run by cell phone so the radio is not tied up.
  C. assigned a separate channel dedicated for use on that scene only.
  D. Both A and C are correct.

_____ **15.** A special communications device which allows the hearing or speech impaired to communicate via telephone is known as a _____ system.
  A. commercial phone
  B. TDD/TTY text phone
  C. direct line
  D. wireless

_____ **16.** The important difference between Basic 911 and Enhanced 911 is that:
  A. enhanced systems have the capability to provide the caller's telephone number and address.
  B. enhanced systems are used only in rural areas.
  C. basic systems are more reliable than enhanced.
  D. basic systems have the capability to provide the caller's telephone number and address.

_____ **17.** Computer-aided dispatch is:
  A. a computer-based, automated system that assists the telecommunicator in assessing dispatch information and recommends responses.
  B. an organized collection of similar facts.
  C. typically used by operations chief officers in the fire service.
  D. an emergency alerting device primarily used by volunteer department personnel to receive reports of emergency incidents.

_____ **18.** Which of the following **is** **not** a polar solvent?
  A. Alcohol
  B. Acetone
  C. Kerosene
  D. Ketone

_____ **19.** For application of aqueous film-forming foam, educators or proportioners operate on a/an _____ principle.
  A. static pressure
  B. venturi
  C. induction
  D. positive pressure

_____ **20.** Solid bore or solid tip nozzles can be effectively used for foam application with:
   A. Aqueous Film Forming Foam (AFFF).
   B. protein foam.
   C. Compressed Air Foam Systems (CAFS).
   D. Film-Forming Fluoroprotein (FFFP).

_____ **21.** Which of the following **is not** one of the ways foam is applied using a nozzle?
   A. Raindown or snowflake technique
   B. Bank-in or roll-on technique
   C. Sub-surface injection technique
   D. Bank-down or off the wall technique

_____ **22.** A roof that is elevated in the center and with an angular slope to the edges is called a _____ roof.
   A. butterfly
   B. dome
   C. pitched/gabled
   D. double-angle

_____ **23.** Spalling of concrete could lead to early collapse in Type I buildings because:
   A. loss of moisture in concrete reduces its fire rating.
   B. the added weight of broken pieces may cause overload.
   C. it could create void spaces.
   D. reinforcing steel is exposed to the heat of the fire.

_____ **24.** Fire fighters should know that fire in Type V construction presents:
   A. shortening of steel components.
   B. breakdown of the concrete members due to the heat buildup.
   C. extensive spalling.
   D. high potential for fire extension within the building.

_____ **25.** Which of the following is a hazard associated with truss and lightweight construction?
   A. If one member fails, the entire truss is likely to fail.
   B. Once a truss fails, the one next to it is likely to fail.
   C. Trusses will begin to fail after a short period of time of exposure to fire.
   D. All of the above.

_____ **26.** One thing common to all types of trusses is that if one member fails:
   A. only that member will fail.
   B. the truss next to it will keep it from failing completely.
   C. the entire truss is likely to fail.
   D. it is unlikely to have a total collapse.

_____ **27.** **Directions**: Read the following statements, then select your answer from alternatives A–D below.

To ensure that there is little danger of injury, a fire ax should be carried:

1. on the shoulder with the edge pointed toward the ground.

2. with the ax blade away from the body, or protected.

3. with pick-head axes, the pick should be covered with a hand.

    A. All three statements are true.

    B. Statements 1 and 2 are false; statement 3 is true.

    C. Statement 1 is true; statements 2 and 3 are false.

    D. Statement 1 is false; statements 2 and 3 are true.

_____ **28.** Which of the following is a form of electrical heat energy?

    A. Nuclear

    B. Chemical

    C. Flashover

    D. Arcing

_____ **29.** Natural roof ventilation openings consist of:

    A. cutting a 4 ft × 4 ft hole.

    B. trench cutting and smoke ejectors.

    C. scuttle hatches, skylights, stairwell openings and bulkheads.

    D. fire streams from aerial ladders directed across a ventilation opening.

_____ **30.** Which of the following is considered a natural or normal roof opening?

    A. Parapet

    B. Skylight

    C. Soffit

    D. Fire stop

_____ **31.** Before cutting an opening in a roof, fire fighters should:

    A. inspect their cutting tools for sharpness.

    B. check for natural or existing openings.

    C. open all top windows on the windward side of the building.

    D. open all bottom windows on the leeward side of the building.

_____ **32.** To localize a fire and create a chimney effect, the primary ventilation hole should be placed:

    A. at the far end of the roof on the leeward side.

    B. directly over the fire.

    C. at the far end of the roof on the windward side.

    D. in an area where supplemental forced or mechanical ventilation can easily be added later, if needed.

_____ **33.** In high-rise fire fighting situations, typically the fire attack will be initiated from:
   A. the fire floor.
   B. one floor below the fire.
   C. one floor above the fire.
   D. two floors below the fire.

_____ **34.** Construction featuring exterior walls and structural members that are noncombustible or of limited combustible materials without additional fire-resistant protection is Type _____ construction.
   A. I
   B. II
   C. III
   D. V

_____ **35.** During a fire, the control valve on a sprinkler system should be closed:
   A. as soon as the fire department arrives.
   B. after ensuring the fire is out or completely under control.
   C. prior to advancing hose lines into fire area.
   D. when building occupants decide fire is out.

_____ **36.** <u>Directions</u>: Read the following statements regarding reinforced concrete as a building construction material and select your answer from choices A–D.
   1. Concrete is usually reinforced to increase its compressive strength.
   2. Fire damage to reinforced concrete **primarily** consists of spalling.
   3. Spalling of reinforced concrete can lead to building collapse.

   A. Statement 1 is true; statements 2 and 3 are false.
   B. Statements 1 and 2 are true; statement 3 is false.
   C. Statement 1 is false; statements 2 and 3 are true.
   D. All three statements are true.

_____ **37.** A report that is made to the Incident Commander signifying that companies working in the hazard zone are all safe and accounted is called:
   A. the all clear signal.
   B. PAR.
   C. loss is stopped.
   D. status report.

_____ **38.** All of the following are **true** regarding the use of a fog stream on an interior fire attack **except**:
   A. discharge pattern can be adjusted to suit the situation.
   B. is not affected by wind.
   C. can aid in ventilation.
   D. provide maximum protection to fire fighters.

_____ **39.** Pressurized flammable liquids and gases should:

    A. always be extinguished.

    B. <u>not</u> be extinguished unless the fuel can be immediately shut off.

    C. <u>not</u> be extinguished by fire fighters; trained specialists should be called for these fires.

    D. <u>not</u> be extinguished unless the product involved has a vapor density greater than one.

_____ **40.** A BLEVE:

    A. most commonly occurs when flames contact the relief valve.

    B. can occur when insufficient water is applied to keep the tank cool.

    C. is a slow deterioration of the tank.

    D. is a condition caused by consolidation of vaporization.

_____ **41.** LPG is _____ than air.

    A. 1.5 times lighter

    B. 2 times lighter

    C. 1.5 times heavier

    D. 0.5 times heavier

_____ **42.** Prior to the fire investigator's arrival, fire fighters should _____ any evidence found.

    A. tag and photograph

    B. protect and preserve

    C. collect and package

    D. isolate and remove

_____ **43.** Initial recognition and preservation of evidence is the responsibility of the:

    A. fire marshal.

    B. fire inspector.

    C. fire investigator.

    D. fire fighter.

_____ **44.** <u>Directions</u>: Read the following statements regarding discovery and preservation of evidence and select your answer from choices A–D.

    1. Fire fighters have a responsibility to preserve evidence that could indicate the origin or cause of fire.

    2. Evidence discovered should be left in place and protected, and a Company Officer or the fire investigator should be notified immediately.

    3. Evidence is most often found during salvage and overhaul.

        A. Statement 1 is true, statements 2 and 3 are false.

        B. Statements 1 and 2 are true, statement 3 is false.

        C. Statement 1 is false; statements 2 and 3 are true.

        D. All three statements are true.

_____ **45.** Hybrid vehicles present the following hazards to rescuers:
  A. air bags can self-explode while the power is on.
  B. high-voltage cables that have an electrocution hazard.
  C. fuel tanks in the rear that can lead to explosion hazards.
  D. fuel sources have a tendency to leak.

_____ **46.** If possible, entry to a vehicle for rescue purposes should be made through the:
  A. windshield.
  B. doors.
  C. roof.
  D. trunk/hatch.

_____ **47.** A spring-loaded center punch is a good tool to use for breaking _____ vehicle glass.
  A. laminated
  B. tempered
  C. lexan
  D. plate

_____ **48.** Which of the following statements regarding air bags is **incorrect**?
  A. Bags must be on or against a solid base.
  B. You must crib as you lift.
  C. Do **not** stack more than two bags.
  D. When stacking two bags of different sizes the larger bag goes on top.

_____ **49.** In doing a dash displacement, the dash is pushed forward by either a mechanical high lift jack or:
  A. a medium pressure airbag.
  B. a high pressure airbag.
  C. a hydraulic ram.
  D. an air chisel.

_____ **50.** Of the situations listed below, which **would not** apply to a shoring situation?
  A. Vehicle chocking
  B. Earth openings
  C. Cave-ins
  D. Building collapse

_____ **51.** Powered hydraulic tools open and close by use of:
  A. air.
  B. fluid.
  C. steam.
  D. mechanical advantage.

_____ **52.** When using the blanket drag, patients should always be dragged:
  A. feet first.
  B. head first.
  C. sideways.
  D. Either B or C is correct.

_____ **53.** While conducting an inspection, a pre-planning site visit or survey, the fire fighter should check all of the following **except**:
  A. ensure that fire protection equipment, such as fire alarm pull stations and manual activation controls for exhaust hood suppression systems, are accessible.
  B. that the building fire alarm system is functional by simply activating a pull station.
  C. that the standpipe system connections and Fire Department Connections are free of obstructions.
  D. ensure that all sprinkler control valves are open and readily accessible, and that the Fire Department Connection is free of obstructions.

_____ **54.** Building construction is an important factor to be identified in a preincident survey. While conducting a survey on a property, you observe the structure members are of noncombustible materials but may not have fire resistive protection. How would you identify this construction type on the survey form?
  A. Type I
  B. Type II
  C. Type III
  D. Type IV

_____ **55.** Purposes for fire company surveys include all of the following **except** to:
  A. detect and eliminate hazards.
  B. collect information for prefire planning.
  C. provide a show of force to the public and building owner.
  D. provide valuable life safety information services to property owners.

_____ **56.** A facility in which there is a great potential likelihood of life or property loss from a fire is called a/an _____ hazard.
  A. special
  B. assembly
  C. target
  D. industrial

_____ **57.** What is a voluntary home fire inspection that is requested by a homeowner called?
  A. A G-code inspection
  B. A secondary code inspection
  C. A home fire safety survey
  D. A Class II inspection

_____ **58.** One of the goals of public fire safety education is to:
  A. wake people up to the legal ramifications of having a fire.
  B. awaken public interest and support in the fire department.
  C. teach people how to react if a fire occurs.
  D. lower insurance rates.

_____ **59.** Which of the following statements regarding preparing for an inspection visit is **<u>incorrect</u>**?
  A. Plan the area to be inspected.
  B. Review occupancy files prior to leaving the station.
  C. Inspections are performed around firefighting schedules and not the schedule of the business owner.
  D. Give consideration to the type of activities conducted at the business relative to the time of day chosen for the inspection.

_____ **60.** The simplest and **<u>most</u> <u>effective</u>** method of achieving the fire service goal of the preservation of life and property is:
  A. prevention.
  B. improved technology.
  C. more fire fighters.
  D. more fire stations.

_____ **61.** One topic a fire fighter may be asked to present during a fire safety presentation to an external group is:
  A. fire stream applications.
  B. home safety practices.
  C. fire extinguisher maintenance.
  D. rescue practices.

_____ **62.** The first step in making a fire safety presentation is to:
  A. transfer facts and ideas.
  B. practice new ideas.
  C. explain information.
  D. prepare the audience for learning.

_____ **63.** Manual detection systems typically have two problems. The first is that many are local only, and the second is that:
  A. they are very technical to use.
  B. they are placed in closets to reduce false alarms.
  C. building occupants need a key to operate system.
  D. a person must be present to activate the system.

_____ **64.** Which of the following statements is **<u>incorrect</u>**?
  A. Sprinkler systems are designed to automatically distribute water through sprinklers that are placed at set intervals on a system of piping to extinguish or control the spread of fires.
  B. Most sprinkler heads detect the heat of a fire and begin to apply water directly over the source of the heat.
  C. Sprinkler systems are ineffective.
  D. Sprinkler heads, unless deluge type heads, are heat sensitive-devices that react to a fixed temperature.

_____ **65.** The _____ system is equipped with all sprinkler heads of the open type.
  A. wet-pipe
  B. dry-pipe
  C. deluge
  D. preaction

_____ **66.** A fire department connection to a sprinkler system enables fire fighters to:
  A. connect hand lines for attacking the fire.
  B. drain water from the system.
  C. increase water supply.
  D. test the system.

_____ **67.** Which of the following **should** **not** be used as a control valve for an automatic sprinkler system?
  A. Post indicator valve
  B. Wall post indicator valve
  C. Outside screw and yoke
  D. Gate valve

_____ **68.** Residential sprinkler systems are wet- or dry-pipe systems. The recommended piping for these systems is constructed of:
  A. lead.
  B. ductile iron.
  C. brass.
  D. steel or plastic.

_____ **69.** Which of the following sprinkler system components is used to limit water flow to one direction?
  A. OS&Y valve
  B. Check valve
  C. Control valve
  D. Butterfly valve

_____ **70.** During the preincident site visit/survey, what information should be obtained and documented?
  A. Built-in fire protection
  B. Construction type
  C. Structure size, height and number of stories
  D. All of the above.

_____ **71.** The full name of an OS&Y valve is _____ valve.
  A. open stem and yoke
  B. outside, shut, and yoke
  C. outside shield and yoke
  D. outside stem and yoke

_____ **72.** The types of valves found in water supply distribution systems are:
    A. indicating and non-indicating.
    B. ball and check.
    C. in-line and flow.
    D. all of the above.

_____ **73.** Which of the following **is** **not** one of the component parts of a dry-barrel fire hydrant?
    A. Operating stem
    B. Stem nut
    C. Post-indicator gate
    D. Drain hole

_____ **74.** Which of the following **is** **not** a component of a grid system?
    A. Primary feeders
    B. Secondary feeders
    C. Distributors
    D. Risers

_____ **75.** The following illustration depicts a _____ hydrant.
    A. dry-barrel
    B. wet-barrel
    C. drafting
    D. dry

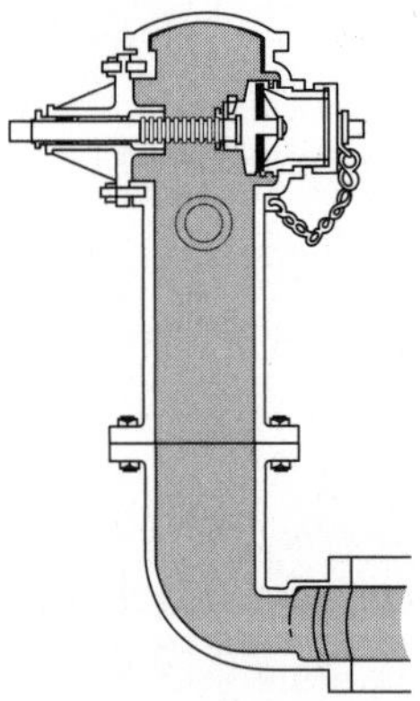

_____ **76.** Which of the following **is** **not** one of the three common systems for water system distribution?
    A. Artesian
    B. Direct pumping
    C. Gravity
    D. Combination

_____ **77.** The four fundamental components of a modern water system are:
    A. source, mains, feeders, and risers.
    B. primary, secondary, standpipes, and subscriber connections.
    C. pipes, valves, hydrants, and pumps.
    D. source, means of moving, treatment plant, and distribution system.

_____ **78.** A fire hydrant that receives water from only one direction is called
a _____ hydrant.
A. one-way
B. steamer
C. circulating-feed
D. dead-end

_____ **79.** The smaller internal grid arrangement of a water distribution system that feeds
hydrants, as well as the domestic and commercial requirements, **best** describes:
A. primary feeders.
B. secondary feeders.
C. distributors.
D. grid network.

_____ **80.** Large pipes that carry large quantities of water to various points along the
water supply system for distribution to smaller mains **best** defines:
A. primary feeders.
B. secondary feeders.
C. distributors.
D. grid network.

_____ **81.** The following illustration depicts a _____ hydrant.
A. dry-barrel
B. wet-barrel
C. drafting
D. dry

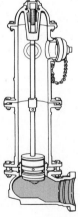

_____ **82.** A network of intermediate-sized pipe that reinforces the overall grid system by
forming loops that interlock primary feeders **best** defines:
A. primary loop.
B. secondary feeders.
C. distributors.
D. grid network.

_____ **83.** Which statement regarding residential sprinklers is <u>incorrect</u>?

    A. They are smaller versions of wet- or dry-pipe sprinkler systems.

    B. The water supply is combined with the domestic system.

    C. The use of plastic pipe is allowed.

    D. They are designed to control the level of fire involvement such that residents can escape.

_____ **84.** How does a deluge system differ from other types of sprinkler systems?

    A. It has air in the system until activated.

    B. It is installed in floors not ceilings.

    C. All sprinkler heads are open.

    D. All of the above.

_____ **85.** A post indicator valve (PIV) is:

    A. a device to speed the operation of the dry-pipe valve by detecting the decrease in air pressure.

    B. designed to control the head pressure at the outlet of a standpipe system to prevent excessive nozzle pressures in hose lines.

    C. a control valve that is mounted on a post case with a small window reading either "open" or "shut."

    D. a control valve that is mounted on a wall in a metal case with a small window reading either "open" or "shut."

_____ **86.** Dry-pipe systems are used in all of the following incidents <u>except</u>:

    A. in buildings that refrigerate or freeze materials.

    B. in unheated buildings.

    C. outdoor applications where freezing temperatures occur.

    D. where rapid activation is required.

_____ **87.** Failure to open a dry-barrel hydrant fully will result in a reduced amount of available water and will contribute to:

    A. sedimentation.

    B. susceptibility to freezing.

    C. difficulty in draining the main.

    D. ground erosion.

_____ **88.** <u>Directions</u>: Read the following statements regarding fire hydrants and select your answer from Choices A–D.

    1. The valve that controls the flow of water in a dry-barrel hydrant is located below ground and the frost line.

    2. The drain opening found at the bottom of the barrel is an exclusive feature of dry-barrel hydrants.

    3. A dry-barrel hydrant should always be either fully open or fully closed.

        A. Statement 1 is true; statements 2 and 3 are false.

        B. Statements 1 and 2 are true; statement 3 is false.

        C. Statement 1 is false; statements 2 and 3 are true.

        D. All three statements are true.

_____ **89.** <u>Directions</u>: Read the following statements regarding fire hydrants and select your answer from choices A–D.

    1. Mechanical damage to fire hydrants can be caused by many things including nature, vandals, accidents, and improper actions by members of the fire department.

    2. Rust and corrosion of hydrants should be noted for referral to the water department during hydrant inspection by fire fighters.

    3. Sedimentation and encrustation can restrict or completely obstruct flow from hydrants.

    A. Statement 1 is true; statements 2 and 3 are false.

    B. Statements 1 and 2 are true; statement 3 is false.

    C. Statement 1 is false; statements 2 and 3 are true.

    D. All three statements are true.

_____ **90.** The simplest sprinkler system in design and operation is the _____ system.

    A. wet-pipe

    B. deluge

    C. dry-pipe

    D. preaction

_____ **91.** The full name of a PIV is _____ valve.

    A. position-indicator

    B. post-indicator

    C. point-indicator

    D. positive-indicator

_____ **92.** _____ systems remain the most reliable of all fire protection devices.

    A. Foam

    B. Dry chemical

    C. Automatic sprinkler

    D. Standpipe

_____ **93.** Under normal circumstances, the air pressure gauge on a dry-pipe sprinkler system will read _____ the water pressure gauge.

    A. the same as

    B. higher than

    C. lower than

    D. almost double

_____ **94.** Which of the following <u>is</u> <u>**not**</u> a type of sprinkler system?

    A. Preaction

    B. Dry-pipe

    C. Deluge

    D. Total flooding

_____ **95.** When shutting down a wet-pipe sprinkler system, one should first turn off the main water control valve and:
A. disable the alarm check valve.
B. open the exhauster to drain the system.
C. drain the system.
D. close all riser indicating valves.

_____ **96.** When filling a hose line with water for testing, a pump pressure of approximately _____ psi is maintained.
A. 25 to 30
B. 45 to 50
C. 80 to 90
D. 250

_____ **97.** When conducting a hose service test, after charging the hose line with water:
A. ensure all air has been discharged from the line.
B. have an adequate number of fire fighters controlling each line.
C. tighten each coupling as tightly as possible.
D. walk each line to determine whether there are any air leaks.

_____ **98.** After the proper test pressure has been reached during an annual service test on fire hose, pressure should be maintained for a period of:
A. thirty seconds.
B. one to three minutes.
C. three to five minutes.
D. twenty minutes.

_____ **99.** The **maximum** length of time that fire hose should be used without a service test is:
A. six months.
B. one year.
C. three years.
D. five years.

_____**100.** When laying out fire hose to be service tested, test lengths should be:
A. no more than 150 feet.
B. no more than 300 feet.
C. no more than 500 feet.
D. no more than 200 feet.

Did you score higher than 80 percent on Examination II-1? Circle Yes or No in ink.

Now that you have finished the feedback step for Examination II-1, it is time to repeat the process by taking another comprehensive examination for NFPA 1001.

Examination II-2, Adding Difficulty and Depth

During Examination II-2, progress will be made in developing your depth of knowledge and skills. Reminder: Follow the steps carefully to realize the best return on effort.

Step 1—Take Examination II-2. When you have completed Examination II-2, go to Appendix B and compare your answers with the correct answers.

Step 2—Score Examination II-2. How many examination items did you miss? Write the number of missed examination items in the blank in ink _____. Enter the number of examination items you guessed in this blank _____. Enter these numbers in the designated locations on your Personal Progress Plotter.

Step 3—Once again, the learning begins. During the feedback step, research the correct answer using the Appendix B information for Examination II-2. Highlight the correct answer during your research of the reference materials. Read the entire paragraph containing the correct answer.

Examination II-2

Directions

Remove Examination II-2 from the manual. First, take a careful look at the examination. There should be 100 examination items. Notice that a blank line precedes each examination item number. This line is provided for you to enter the answer to the examination item. Write the answer in ink. Remember the rule about not changing your answers. Our research shows that changed answers are most often changed to an incorrect answer, and, more often than not, the answer that is chosen first is correct.

If you guess the answer to a question, place an "X" or a check mark by your answer. This step is vitally important as you gain and master knowledge. We will explain how we treat the "guessed" items later in SAEP.

Take the examination. Once you complete it, go to Appendix B and score your examination. Once the examination is scored, carefully follow the directions for feedback on the missed and guessed examination items.

_____ **1.** Which of the following **is** **not** a major functional component of the Incident Command System?
  A. Groups
  B. Planning
  C. Logistics
  D. Operations

_____ **2.** The Incident Command System should:
  A. be fully implemented for all situations.
  B. be initiated by the first fire unit on the scene.
  C. provide procedures that perfectly fit all departments.
  D. eliminate the need for mutual aid assistance.

_____ **3.** What is the optimal number of individuals that one person should be supervising at an emergency incident?
  A. Up to two
  B. Eight to ten
  C. Three to seven
  D. Eight or more

_____ **4.** Under the Incident Command System, the _____ Officer is responsible for providing factual and accurate information to the media.
  A. Safety
  B. Liaison
  C. Staffing
  D. Public Information

_____ **5.** A situation status progress report is provided upon:
  A. termination of the incident.
  B. arrival.
  C. transfer of command.
  D. setup of the staging area.

_____ **6.** <u>Directions</u>: Read the following statements regarding transfer of command and select your answer from choices A–D.
   1. The first arriving fire department member must be prepared to transfer command to the next arriving individual with a higher authority.
   2. Transfer of command must include a situation status progress report.
   3. Command can only be transferred to someone who is on scene.

   A. Statement 1 is true, statements 2 and 3 are false.
   B. Statements 1 and 2 are true, statement 3 is false.
   C. Statements 1 and 2 are false, statement 3 is true.
   D. All three statements are true.

_____ **7.** For what types and sizes of incidents is the Incident Command System (ICS) designed?
   A. Multi-agency only, medium or large size
   B. Multi-agency only, any size
   C. Single agency only, large size
   D. All types and sizes

_____ **8.** Who is the Incident Commander's point of contact for representatives from outside agencies?
   A. The Planning Director
   B. The Communications Center
   C. The Liaison Officer
   D. The Staging Chief

_____ **9.** By what title or rank are the heads of the four major functional components of the Incident Command System known?
   A. Director
   B. Chief
   C. Manager
   D. Officer

_____ **10.** What Incident Command System term can refer to companies or crews that have been assigned on the basis of either geography or function?
   A. Division
   B. Strike force
   C. Recon
   D. Branch

_____ **11.** The term used for areas adjacent to a building is:
   A. rotundas.
   B. exposures.
   C. sectors.
   D. stages.

_____ **12.** What would be the incident command designation for the 7th floor on a
high-rise?
A. Sector 7
B. Sixth floor
C. Division 7
D. Level 6

_____ **13.** Risk management is:
A. a collection of documents that includes all federally promulgated regulations
for all federal agencies.
B. the process of minimizing the chance, degree, or probability of damage, loss,
or injury.
C. the result of a series of events and conditions that lead to an unsafe situation
resulting in injury and/or property damage.
D. a formal gathering of incident responders to help defuse and address stress
from a given incident.

_____ **14.** Which of the following items **would** **not** be found in the risk/benefit
philosophy of a risk management plan?
A. Where no life can be saved, no risk shall be taken by fire fighters.
B. Situations endangering valued property shall cause fire fighters to take a
calculated and weighted risk.
C. Where no life or valued property can be saved, risk may be taken by
fire fighters.
D. Significant risk to the life of a fire fighter shall be limited to those situations
where the fire fighter can potentially save endangered lives.

_____ **15.** The type of call, action taken, number of injuries or fatalities, and
the property usage information are all entries to be included on
a/an _____ Report.
A. United States Fire Administration
B. Incident Command System
C. Advanced Cardiac Life Support
D. National Fire Incident Reporting System

_____ **16.** The report given to the incident commander from an interior crew which tells
the incident commander that the fire is controlled would be a/an:
A. sizeup/conditions report.
B. situation status/progress report.
C. all clear notification.
D. staging report.

_____ **17.** All of the following are times or events when the Incident Commander may
call for PAR **except** when:
A. initial sizeup is completed.
B. incident is declared under control.
C. there is a change in strategy.
D. there is an emergency evacuation.

_____ **18.** Which one of the following **is not** a method by which foam acts as a suppression agent?
  A. Smothering
  B. Cooling
  C. Inhibiting the chemical chain reaction
  D. Separating

_____ **19.** Firefighting foam solution is _____ percent water.
  A. 95 to 98.6
  B. 80 to 85.5
  C. 94 to 99.9
  D. 80 to 90.7

_____ **20.** The **most effective** type of foam for use on polar solvents is:
  A. alcohol-resistant.
  B. Class A foam.
  C. low/high expansion foam.
  D. FFFP.

_____ **21.** The **preferred** method of controlling flammable liquid fires is:
  A. the use of foam.
  B. the use of large amounts of water.
  C. the use of unmanned nozzles.
  D. letting the fire burn undisturbed.

_____ **22.** **Directions:** Read the following statements regarding foam application and select your answer from choices A–D.
  1. AFFF can be effectively applied through a standard fire department fog nozzle.
  2. Air aspirating foam nozzles or foam nozzle adaptors on standard fog nozzles produce effective, high quality foam.
  3. Special aerating nozzles are recommended for use with protein and fluoroprotein foams.

  A. Statement 1 is true, statements 2 and 3 are false.
  B. Statements 1 and 2 are true, statement 3 is false.
  C. Statements 1 and 2 are false; statement 3 is true.
  D. All three statements are true.

_____ **23.** Foam, rather than water, is chosen to control a hydrocarbon fire because:
  A. it cools more effectively.
  B. it is soluble, which allows it to dilute the fuel concentration.
  C. its specific gravity is greater than that of the burning fuel.
  D. its specific gravity is less than that of the burning fuel.

_____ **24.** The safest recommended means for a fire fighter to disconnect electrical service to a building is to:
  A. cut the service entrance wire.
  B. pull the meter.
  C. locate the nearest transformer and deactivate it.
  D. shut off the main power breaker/fuse in the panel box.

_____ **25.** A roof that is elevated in the center and with an angular slope to the edges is called a _____ roof.
  A. butterfly
  B. dome
  C. pitched/gabled
  D. double-angle

_____ **26.** Hidden fire can be checked by using a/an:
  A. detector for different levels of carbon monoxide and oxygen.
  B. Halligan tool to remove the whole wall.
  C. plaster tool from the opposite side of the wall.
  D. electronic/infrared heat sensor.

_____ **27.** What kind of heat energy is the heat of compression?
  A. Chemical
  B. Electrical
  C. Mechanical
  D. Nuclear

_____ **28.** Positive pressure ventilation is effective:
  A. when opening of doors and windows in the structure can be controlled.
  B. only if you can create a lower pressure zone in the structure.
  C. if the exhaust opening is smaller than the entry point, creating a Venturi effect.
  D. if an entire floor is ventilated at a time, starting at the highest floor and working down.

_____ **29.** Solid streams are preferred whenever:
  A. a large volume of smoke is present.
  B. reach and penetration are needed.
  C. fire fighters need a protective curtain.
  D. forced ventilation is necessary.

_____ **30.** The usual cause of collapse of open-web steel joist is the:
  A. amount of heat generated by the fire in a structure.
  B. poor construction methods.
  C. impact load of fire fighters jumping onto roof.
  D. unknown (still being researched).

_____ **31.** A type of wood framing that has vertical channels going from floor to floor, allowing a fire to travel uninterrupted is a/an _____ frame construction.
   A. platform
   B. open
   C. balloon
   D. box

_____ **32.** Which of the following gases <u>is</u> <u>not</u> produced in fires?
   A. Carbon monoxide
   B. Hydrogen cyanide
   C. Carbon dioxide
   D. Oxygen

_____ **33.** Which of the following statements is <u>incorrect</u>?
   A. The secondary search is the most dangerous.
   B. Searching a building is completed in two different operations—primary and secondary search.
   C. During primary search, the team is often ahead of attack lines and may be above the fire.
   D. The primary search takes place in a rapid but thorough manner in areas most likely to have victims.

_____ **34.** Which of the following statements regarding a trench cut is <u>incorrect</u>?
   A. The trench cut is an offensive action.
   B. The trench cut is from 2–4 feet wide.
   C. It is <u>not</u> opened until the entire cut is complete.
   D. It is made in coordination with interior crews.

_____ **35.** A Rapid Intervention Crew/Team is defined as:
   A. any combination of single resources assembled for an assignment.
   B. the designation for a set number of resources of the same type and kind.
   C. a company designated to search for and rescue trapped fire fighters.
   D. a designated group that is used for rapid knock down of wildland fires.

_____ **36.** During overhaul, fire fighters should wear:
   A. boots and gloves; coats and helmets are unnecessary.
   B. lightweight clothing, due to residual heat.
   C. full protective gear, including SCBA.
   D. full protective gear; SCBA is not needed.

_____ **37.** <u>Directions</u>: Read the following statements regarding masonry as a construction material and select your answer from choices A–D.

1. The term "masonry" applies to brick, block and stone.

2. Masonry is inherently resistive to the effects of fire.

3. Rapid cooling caused by the application of hose streams during fire suppression can cause a masonry wall to spall, crack or otherwise deteriorate.

    A. Statement 1 is true; statements 2 and 3 are false.

    B. Statements 1 and 2 are true; statement 3 is false.

    C. Statement 1 is false; statements 2 and 3 are true.

    D. All three statements are true.

_____ **38.** <u>Directions</u>: Read the following statements regarding steel as a building construction material and select your answers from choices A–D.

1. Steel by itself is fire resistive.

2. Steel will both expand and lose strength as it is heated.

3. There are no accurate indicators that enable fire fighters to predict when a steel beam will fail.

    A. Statement 1 is true; statements 2 and 3 are false.

    B. Statements 1 and 2 are true; statement 3 is false.

    C. Statement 1 is false; statements 2 and 3 are true.

    D. All three statements are true.

_____ **39.** <u>Directions</u>: Read the following statements regarding reinforced concrete as a building construction material and select your answer from choices A–D.

1. Concrete is usually reinforced to increase its compressive strength.

2. Fire damage to reinforced concrete primarily consists of spalling.

3. Spalling of reinforced concrete can lead to building collapse.

    A. Statement 1 is true; statements 2 and 3 are false.

    B. Statements 1 and 2 are true; statement 3 is false.

    C. Statement 1 is false; statements 2 and 3 are true.

    D. All three statements are true.

_____ **40.** An all metal building is classified as a:

    A. Type I.

    B. Type II.

    C. Type III.

    D. Type V.

_____ **41.** The acronym "BLEVE" stands for Boiling Liquid:

    A. Exhausting Vapor Explosion.

    B. Expanding and Venting Explosion.

    C. Expanding Vapor Explosion.

    D. Exhausting Vapor Expansion.

_____ **42.** An increase in the intensity of sound or fire issuing from a relief valve
may indicate:
A. the relief valve is clogged.
B. rupture of the vessel is imminent.
C. the fire is burning out.
D. the tank is cooling down.

_____ **43.** When containers of flammable gases are exposed to flame impingement, the
water for cooling the container should be applied to cool the:
A. container base.
B. relief valve.
C. ends of the tanks.
D. fire fighters.

_____ **44.** It is important for fire fighters responding to reported LPG leaks to remember
that propane has a vapor density of **approximately**:
A. 0.15
B. 1.5
C. 0.6
D. 2.6

_____ **45.** LPG is _____ than air.
A. 1.5 times lighter
B. 2 times lighter
C. 1.5 times heavier
D. 0.5 times heavier

_____ **46.** Which of the following **is** **not** one of the normal observations that fire fighters
make to assist in determining fire cause?
A. How the fire reacted to water application
B. People leaving the fire scene in a hurry
C. Hindrances to firefighting
D. The number and location of observers

_____ **47.** Fire fighters should always remember how they gained entry into a
building because:
A. entry may have an effect on the behavior of the fire.
B. the fire investigator may want to know.
C. you may need to use the opening in case of a rekindle.
D. you should always exit the same way you entered.

_____ **48.** Once a fire investigator has completed the work required
in gathering evidence and information from a fire scene,
a thorough _____ can be done.
A. ventilation
B. demobilization
C. inventory
D. overhaul

_____ **49.** When using a spring-loaded center punch to open a window, the fire fighter should press the center punch at the:
    A. center of the window.
    B. lower corner of the window.
    C. upper corner of the window.
    D. at the top of the window.

_____ **50.** If possible, entry to a vehicle for rescue purposes should be made through the:
    A. windshield.
    B. doors.
    C. roof.
    D. trunk/hatch.

_____ **51.** The phase of vehicle extrication that determines the nature and extent of the overall situation is:
    A. preparation.
    B. disentanglement.
    C. hazard control.
    D. sizeup.

_____ **52.** The key to an efficient extrication operation is proper _____ of the situation.
    A. dispatch
    B. sizeup
    C. triage
    D. assignment

_____ **53.** If rescuers are able to get into a vehicle, their first action should be to:
    A. begin extraction.
    B. place a backboard.
    C. assess/protect the victim(s).
    D. remove broken glass.

_____ **54.** You are operating as a member of a rescue company at the scene of an auto extrication. The scene, vehicle, and the patient(s) are stabilized, all hazards are controlled, and access and disentanglement have been accomplished. The next step in the extrication process should be:
    A. termination and post-incident analysis.
    B. patient packaging and patient removal.
    C. transporting patient(s) to appropriate facilities.
    D. initiating extrication operations on another vehicle, if needed.

_____ **55.** The first power hydraulic tool that became available to fire fighters was:
    A. cutters (also known as shears).
    B. spreaders.
    C. the combination tool.
    D. ram.

_____ **56.** When extricating a patient from a motor vehicle crash, disassemble refers to the:
A. displacement of major parts (i.e., doors, roof, dash).
B. cutting off of components (i.e., brake pedal, steering wheel).
C. bending of sheet metal or components.
D. actual taking apart of the vehicle components.

_____ **57.** After the crash, some vehicles come to rest on their roofs or sides. What is their status regarding stabilization?
A. None is required, they are usually very stable in either position.
B. They require stabilization. Rescue lift airbags by themselves, one on each side, is the best manner of stabilization.
C. They require stabilization. Rescue lift airbags in conjunction with wedges, one on each end, is the best manner of stabilization.
D. Generally, a combination of cribbing, ropes, webbing, and chains are used to accomplish these types of stabilization tasks.

_____ **58.** Which of the following saw types would **most** **likely** be used during vehicle extrication?
A. Keyhole saw
B. Reciprocating saw
C. Chain saw
D. Coping saw

_____ **59.** To de-energize an airbag, what is the **first** step of the two-step procedure?
A. Ground the solenoid.
B. Remove the cartridge.
C. Unscrew the primer.
D. Disconnect the battery.

_____ **60.** The process of erecting materials such as wood panels, timber, or jacks to strengthen a wall or prevent further collapse is known as:
A. shoring.
B. cribbing.
C. packing.
D. supporting.

_____ **61.** The **most** **common** hazard encountered at a confined space incident is:
A. an oxygen deficient atmosphere.
B. engulfment.
C. collapse of structure.
D. entrapment.

_____ **62.** When lifting rescued persons or heavy objects fire fighters should:
A. use their back to lift, not their legs.
B. use their legs to lift, not their back.
C. try to twist and reach at the same time.
D. squat and lift with shoulders and arms.

_____ **63.** The **most** **common** method of establishing the control zones for an emergency incident site is:
   A. stationing fire fighters along a perimeter.
   B. the use of police or fire line tape.
   C. stationing police along a perimeter.
   D. police-style pedestrian traffic barricades.

_____ **64.** Concrete has excellent _____ strength when it cures.
   A. shear
   B. compressive
   C. torsional
   D. tensile

_____ **65.** While conducting an inspection, a pre-planning site visit or survey, the fire fighter should check all of the following **except**:
   A. insure that fire protection equipment, such as fire alarm pull stations and manual activation controls for exhaust hood suppression systems, are accessible.
   B. that the building fire alarm system is functional by simply activating a pull station.
   C. that the standpipe system connections and Fire Department Connections are free of obstructions.
   D. insure that all sprinkler control valves are open and readily accessible, and that the Fire Department Connection is free of obstructions.

_____ **66.** A connecting plate used in truss construction that can be made of flat steel stock, light gauge metal, or plywood is the definition of a:
   A. joint.
   B. gusset plate.
   C. column.
   D. joist.

_____ **67.** Building construction is an important factor to be identified in a preincident survey. While conducting a survey on a property, you observe the structure members are of noncombustible materials but may not have fire resistive protection. How would you identify this construction type on the survey form?
   A. Type I
   B. Type II
   C. Type III
   D. Type IV

_____ **68.** Purposes for fire company surveys include all of the following **except** to:
   A. detect and eliminate hazards.
   B. collect information for prefire planning.
   C. provide a show of force to the public and building owner.
   D. provide valuable life safety information services to property owners.

_____ **69.** What should be done with smoke alarms when conducting a residential fire safety survey?
   A. Nothing should be done in order to avoid showing possible favoritism towards one brand or another.
   B. If smoke detectors are observed, fire fighters should comment favorably.
   C. One detector at random should be checked, while the occupants look on.
   D. Each one should be verified and tested.

_____ **70.** During the preincident site visit, what information should be obtained and documented?
   A. Built-in fire protection
   B. Access points to the site and interior of the structure
   C. Structure size, height, and number of stories
   D. All of the above.

_____ **71.** Which of the following is a factor in deciding to preplan a structure or area?
   A. Type of hazards expected
   B. Complexity of fire fighting operations
   C. Nature of activities conducted at the occupancy
   D. All of the above.

_____ **72.** With regard to portable fire extinguishers, which of the following situations would a fire fighter bring to the occupant/owner's attention?
   A. The extinguisher has the proper classification and rating for its location.
   B. The pressure gauge indicates that the extinguisher isn't properly charged.
   C. The fire extinguisher is hung on the wall and is easily visible and accessible.
   D. The tag indicates that the extinguisher has been serviced to local laws.

_____ **73.** In addition to providing a service, private dwelling inspections also:
   A. afford the opportunity to introduce members of the fire department.
   B. provide an educational and advisory service.
   C. afford an opportunity to impress upon the public the benefits of the fire department.
   D. All of the above.

_____ **74.** The first step in making a fire safety presentation is to:
   A. transfer facts and ideas.
   B. practice new ideas.
   C. explain information.
   D. prepare the audience for learning.

_____ **75.** Heat detectors:
   A. cannot be used as part of a suppression system.
   B. are slow to activate.
   C. are expensive to install and operate.
   D. are responsible for most false alarms.

_____ **76.** Smoke detectors work primarily on the principles of photoelectricity and:
    A. rate of rise.
    B. fixed temperature.
    C. ionization.
    D. laser beam.

_____ **77.** The _____ system is equipped with all sprinkler heads of the open type.
    A. wet-pipe
    B. dry-pipe
    C. deluge
    D. preaction

_____ **78.** Which of the following **should** **not** be used as a control valve for an automatic sprinkler system?
    A. Post indicator valve
    B. Wall post indicator valve
    C. Outside screw and yoke
    D. Gate valve

_____ **79.** An OS&Y valve is open when the stem threads are:
    A. retracted.
    B. extended.
    C. reversed.
    D. crossed.

_____ **80.** A sprinkler deflector converts a water stream into:
    A. spray.
    B. steam.
    C. a fine mist.
    D. a solid stream.

_____ **81.** During the preincident site visit/survey, what information should be obtained and documented?
    A. Built-in fire protection
    B. Construction type
    C. Structure size, height and number of stories
    D. All of the above.

_____ **82.** The _____ process requires fire fighters to become familiar with community structures.
    A. preventive
    B. code enforcement
    C. investigative
    D. survey/site visit

_____ **83.** A fire hazard that arises from processes or operations that are related to a specific occupancy **best** defines a/an _____ hazard.
   A. special
   B. common
   C. expected
   D. target

_____ **84.** Before conducting a preincident survey and inspection for the purpose of prefire planning, the occupant should be notified in advance so:
   A. the occupant can correct any fire hazards before the fire department arrives.
   B. there will be minimum inconvenience to the owner, occupant, and fire department.
   C. the owner can have fire extinguishers filled.
   D. someone will be at the location to fill out fire department forms.

_____ **85.** In a sprinkler system, what is the function of the valve known as the backflow preventer or clapper?
   A. It prevents water from entering the system.
   B. It prevents water from the system from reentering the public water supply.
   C. It prevents water from entering noninvolved areas.
   D. It protects the system from water surges.

_____ **86.** A sprinkler head with a temperature rating of 135°–170°F would be:
   A. orange.
   B. red.
   C. white.
   D. either uncolored or black.

_____ **87.** Which of the following **is not** considered to be an alarm initiating device?
   A. Water detector
   B. Visible products-of-combustion detector
   C. Flame detector
   D. Invisible products-of-combustion detector

_____ **88.** Which type of alarm initiating device contains a small amount of radioactive material in its sensing chamber?
   A. Water detector
   B. Visible products-of-combustion detector
   C. Flame detector
   D. Ionization detector

_____ **89.** Upon receiving an alarm from a proprietary alarm system, the personnel manning the central alarm point should:
   A. dispatch fire apparatus.
   B. leave their stations and investigate.
   C. reset the alarm.
   D. contact the appropriate agency.

_____ **90.** A local protective signaling system is intended primarily to alert:
   A. the fire department dispatcher.
   B. a private alarm contractor.
   C. occupants of the protected area.
   D. the hotel desk clerk.

_____ **91.** Visible products of combustion are **best** detected by
   a/an _____ detector.
   A. photoelectric
   B. ionization
   C. ultraviolet
   D. infrared

_____ **92.** A/An _____ detector is one that responds to sudden changes
   in temperature.
   A. disk thermostat
   B. photoelectric
   C. ionization
   D. rate-of-rise

_____ **93.** A _____ system is used to protect large commercial and
   industrial buildings and is staffed and supervised on the property.
   A. central station
   B. local protective
   C. remote
   D. proprietary

_____ **94.** Which of the following **is not** a common type of automatic alarm system?
   A. Central Station Protective Signaling System
   B. Local Protective System
   C. Auxiliary Protective Signaling System
   D. Cell Phone Initiated System

_____ **95.** A detector that is usually electrically operated with a bimetallic strip of
   two metals that expand at different rates, eventually bending to touch a
   contact point and complete the alarm circuit is called
   a/an _____ detector.
   A. carbon monoxide
   B. ionization smoke
   C. photoelectric smoke
   D. heat detector

_____ **96.** While preplanning a new business in your run district, you notice a Class K
   fire system. A Class K system be used for:
   A. metal rims.
   B. flammable solids.
   C. cooking oils.
   D. electrical boxes.

_____ **97.** Which of the following **is** **not** one of the commonly used release mechanisms to activate a sprinkler?
   A. Fusible links
   B. Valve cap
   C. Frangible bulb
   D. Chemical pellet

_____ **98.** To determine whether there is any slippage of couplings when testing hose:
   A. measure and mark the exact length of each coupling.
   B. mark the hose by each coupling using a pencil/marker.
   C. check to see whether couplings remain tight.
   D. use a special torque wrench set to 60 psi.

_____ **99.** When testing fire hose, fill each hose with water and, during the filling process, make sure that each nozzle:
   A. receives the same gpm.
   B. is closed and strapped in place.
   C. is held by a fire fighter when discharging water.
   D. is open until all air is discharged.

_____**100.** The type of hose testing conducted by the fire department is called:
   A. acceptance.
   B. pressure.
   C. service.
   D. fireground.

Did you score higher than 80 percent on Examination II-2? Circle Yes or No in ink.

Now that you have finished the feedback step for Examination II-2, it is time to repeat the process by taking another comprehensive examination for NFPA 1001.

Examination II-3, Surveying Weaknesses and Improving Examination-Taking Skills

Examination II-3 is designed to identify your remaining weaknesses in areas covered by NFPA 1001. This examination is randomly generated and contains examination items you have taken before as well as new ones. Some steps in SAEP will require self-study of specific reference materials.

Mark all answers in ink to ensure that no corrections or changes are made later. Do not mark through answers or change answers in any way once you have selected your answer.

Step 1—Take Examination II-3. When you have completed Examination II-3, go to Appendix B and compare your answers with the correct answers.

Step 2—Score Examination II-3. How many examination items did you miss? Write the number of missed examination items in the blank in ink _____. Enter the number of examination items you guessed in this blank _____. Enter these numbers in the designated locations on your Personal Progress Plotter.

Step 3—Now you will begin reinforcing what you have learned! During the feedback step, research the correct answer using the Appendix B information for Examination II-3. Highlight the correct answer during your research of the reference materials. Read the entire paragraph containing the correct answer.

Examination II-3

Directions

Remove Examination II-3 from the manual. First, take a careful look at the examination. There should be 100 examination items. Notice that a blank line precedes each examination item number. This line is provided for you to enter the answer to the examination item. Write the answer in ink. Remember the rule about not changing your answers. Our research shows that changed answers are most often changed to an incorrect answer, and, more often than not, the answer that is chosen first is correct.

If you guess the answer to a question, place an "X" or check mark by your answer. This step is vitally important as you gain and master knowledge. We will explain how we treat the "guessed" items later in SAEP.

Take the examination. Once you complete it, go to Appendix B and score your examination. Once the examination is scored, carefully follow the directions for feedback on the missed and guessed examination items.

_____ **1.** Policies are examples of standing plans designed to provide:
   A. staffing requirement guidelines.
   B. guidance for decision making.
   C. problem-solving.
   D. communications.

_____ **2.** A procedure is a/an:
   A. guide to thinking.
   B. detailed guide to action.
   C. guide to decision making.
   D. interpretation.

_____ **3.** Which of the following is one of the major staff functions/components of the Incident Command System?
   A. PIO
   B. Planning
   C. Liaison
   D. Safety officer

_____ **4.** Which of the following <u>is</u> <u>**not**</u> a characteristic of the Incident Command System?
   A. Integrated communications
   B. Predesignated facilities
   C. Modular organization
   D. Independent action plans

_____ **5.** The Incident Command System should:
   A. be fully implemented for all situations.
   B. be initiated by the first fire unit on the scene.
   C. provide procedures that perfectly fit all departments.
   D. eliminate the need for mutual aid assistance.

_____ **6.** The tracking of personnel working at an incident requires a system that is standardized for every incident to establish:
A. accountability.
B. chain of command.
C. unity of command.
D. span of control.

_____ **7.** In the Incident Command System, _____ are functional crews assigned to perform a specific task such as ventilation or rescue.
A. divisions
B. groups
C. single resources
D. branches

_____ **8.** Within the Incident Command System, the positions of Safety, Liaison, and Information are:
A. divisions.
B. groups.
C. functional areas.
D. command staff positions.

_____ **9.** The Incident Command System is **best** defined as:
A. the overall plan developed and used to control an incident.
B. an organized, systematic method for the command, control, and management of an emergency incident.
C. the ability to start small and expand if an incident becomes more complex.
D. one designated leader or officer to command an incident.

_____ **10.** Under the Incident Command System, the _____ Officer is responsible for providing factual and accurate information to the media.
A. Safety
B. Liaison
C. Staffing
D. Public Information

_____ **11.** By what title or rank are the heads of the four major functional components of the Incident Command System known?
A. Director
B. Chief
C. Manager
D. Officer

_____ **12.** Which Incident Command System section is responsible for the management of all actions that are directly related to controlling the incident?
A. Logistics
B. Operations
C. Tactics
D. Planning

_____ **13.** Which Incident Command System section is responsible for keeping vehicles fueled and providing food for fire fighters?
A. Logistics
B. Liaison
C. Operations
D. Safety

_____ **14.** What Incident Command System term can refer to companies or crews that have been assigned on the basis of either geography or function?
A. Division
B. Strike force
C. Recon
D. Branch

_____ **15.** As more companies arrive at an escalating incident, what is one reason the command structure must expand?
A. To employ the arriving officers
B. To maintain span of control
C. To allow unity of command
D. To counter the formation of sectorization

_____ **16.** The arrival report should contain:
A. a situation evaluation.
B. the attack mode selected.
C. the person in command.
D. All of the above.

_____ **17.** The two **most common** ways the incident commander orders fire fighters to evacuate a structure are to broadcast a radio message and:
A. page all fire fighters to respond.
B. implement an accountability system.
C. contact dispatch to activate PASS device.
D. sound an audible warning.

_____ **18.** A special communications device which allows the hearing or speech impaired to communicate via telephone is known as a _____ system.
A. commercial phone
B. TDD/TTY text phone
C. direct line
D. wireless

_____ **19.** The important difference between Basic 911 and Enhanced 911 is that:
A. enhanced systems have the capability to provide the caller's telephone number and address.
B. enhanced systems are used only in rural areas.
C. basic systems are more reliable than enhanced.
D. basic systems have the capability to provide the caller's telephone number and address.

_____ **20.** Computer-aided dispatch is:
  A. a computer-based, automated system that assists the telecommunicator in assessing dispatch information and recommends responses.
  B. an organized collection of similar facts.
  C. typically used by operations chief officers in the fire service.
  D. an emergency alerting device primarily used by volunteer department personnel to receive reports of emergency incidents.

_____ **21.** The report given to the incident commander from an interior crew which tells the incident commander that the fire is controlled would be a/an:
  A. sizeup/conditions report.
  B. situation status/progress report.
  C. all clear notification.
  D. staging report.

_____ **22.** All of the following are times or events when the Incident Commander may call for PAR **except** when:
  A. initial sizeup is completed.
  B. incident is declared under control.
  C. there is a change in strategy.
  D. there is an emergency evacuation.

_____ **23.** Class A foams are essentially wetting agents that _____ of water.
  A. increase the viscosity
  B. increase the resistance
  C. reduce the surface tension
  D. create a higher vaporization point

_____ **24.** Firefighting foam solution is _____ percent water.
  A. 95 to 98.6
  B. 80 to 85.5
  C. 94 to 99.9
  D. 80 to 90.7

_____ **25.** Which of the following **is not** one of the ways foam is applied using a nozzle?
  A. Raindown or snowflake technique
  B. Bank-in or roll-on technique
  C. Sub-surface injection technique
  D. Bank-down or off the wall technique

_____ **26.** The **preferred** method of controlling flammable liquid fires is:
  A. the use of foam.
  B. the use of large amounts of water.
  C. the use of unmanned nozzles.
  D. letting the fire burn undisturbed.

_____ **27.** Foam, rather than water, is chosen to control a hydrocarbon fire because:
  A. it cools more effectively.
  B. it is soluble, which allows it to dilute the fuel concentration.
  C. its specific gravity is greater than that of the burning fuel.
  D. its specific gravity is less than that of the burning fuel.

_____ **28.** The safest recommended means for a fire fighter to disconnect electrical service to a building is to:
    A. cut the service entrance wire.
    B. pull the meter.
    C. locate the nearest transformer and deactivate it.
    D. shut off the main power breaker/fuse in the panel box.

_____ **29.** What type of construction has structural members (including walls, columns, beams, floors, and roofs) that are made of noncombustible or limited-combustible materials?
    A. Type I
    B. Type III
    C. Type IV
    D. Type V

_____ **30.** Fire fighters should know that fire in Type V construction presents:
    A. shortening of steel components.
    B. breakdown of the concrete members due to the heat buildup.
    C. extensive spalling.
    D. high potential for fire extension within the building.

_____ **31.** What type of building construction is made up of solid heavy timber or laminated wood?
    A. Type I
    B. Type II
    C. Type IV
    D. Type V

_____ **32.** **Directions:** Read the following statements, then select your answer from alternatives A–D below.
To ensure that there is little danger of injury, a fire ax should be carried:
    1. on the shoulder with the edge pointed toward the ground.
    2. with the ax blade away from the body, or protected.
    3. with pick-head axes, the pick should be covered with a hand.

    A. All three statements are true.
    B. Statements 1 and 2 are false; statement 3 is true.
    C. Statement 1 is true; statements 2 and 3 are false.
    D. Statement 1 is false; statements 2 and 3 are true.

_____ **33.** Hidden fire can be checked by using a/an:
    A. detector for different levels of carbon monoxide and oxygen.
    B. Halligan tool to remove the whole wall.
    C. plaster tool from the opposite side of the wall.
    D. electronic/infrared heat sensor.

_____ **34.** What kind of heat energy is the heat of compression?
    A. Chemical
    B. Electrical
    C. Mechanical
    D. Nuclear

_____ **35.** To localize a fire and create a chimney effect, the primary ventilation hole should be placed:
  A. at the far end of the roof on the leeward side.
  B. directly over the fire.
  C. at the far end of the roof on the windward side.
  D. in an area where supplemental forced or mechanical ventilation can easily be added later, if needed.

_____ **36.** Positive pressure ventilation is effective:
  A. when opening of doors and windows in the structure can be controlled.
  B. only if you can create a lower pressure zone in the structure.
  C. if the exhaust opening is smaller than the entry point, creating a Venturi effect.
  D. if an entire floor is ventilated at a time, starting at the highest floor and working down.

_____ **37.** During a fire, the control valve on a sprinkler system should be closed:
  A. as soon as the fire department arrives.
  B. after ensuring the fire is out or completely under control.
  C. prior to advancing hoselines into fire area.
  D. when building occupants decide fire is out.

_____ **38.** A forward staging area for high-rise fires is usually established _____ floor(s) below the fire floor.
  A. 1
  B. 2
  C. 3
  D. 4

_____ **39.** At a **minimum**, fire fighters must work in teams of _____ when entering an involved structure?
  A. two
  B. three
  C. seven
  D. five

_____ **40.** Which of the following statements regarding a trench cut is **incorrect**?
  A. The trench cut is an offensive action.
  B. The trench cut is from 2–4 feet wide.
  C. It is **not** opened until the entire cut is complete.
  D. It is made in coordination with interior crews.

_____ **41.** During overhaul, fire fighters should wear:
  A. boots and gloves; coats and helmets are unnecessary.
  B. lightweight clothing, due to residual heat.
  C. full protective gear, including SCBA.
  D. full protective gear; SCBA is **not** needed.

_____ **42.** <u>Directions</u>: Read the following statements regarding masonry as a construction material and select your answer from choices A–D.
1. The term "masonry" applies to brick, block and stone.
2. Masonry is inherently resistive to the effects of fire.
3. Rapid cooling caused by the application of hose streams during fire suppression can cause a masonry wall to spall, crack or otherwise deteriorate.

    A. Statement 1 is true; statements 2 and 3 are false.
    B. Statements 1 and 2 are true; statement 3 is false.
    C. Statement 1 is false; statements 2 and 3 are true.
    D. All three statements are true.

_____ **43.** A BLEVE:
    A. most commonly occurs when flames contact the relief valve.
    B. can occur when insufficient water is applied to keep the tank cool.
    C. is a slow deterioration of the tank.
    D. is a condition caused by consolidation of vaporization.

_____ **44.** It is important for fire fighters responding to reported LPG leaks to remember that propane has a vapor density of **approximately**:
    A. 0.15
    B. 1.5
    C. 0.6
    D. 2.6

_____ **45.** LPG is _____ than air.
    A. 1.5 times lighter
    B. 2 times lighter
    C. 1.5 times heavier
    D. 0.5 times heavier

_____ **46.** Which of the following <u>is</u> **<u>not</u>** one of the normal observations that fire fighters make to assist in determining fire cause?
    A. How the fire reacted to water application
    B. People leaving the fire scene in a hurry
    C. Hindrances to firefighting
    D. The number and location of observers

_____ **47.** Following a fire of suspicious origin, the fire fighter should:
    A. carefully wash all burned articles with water to clean them off.
    B. remove any possible evidence to the outdoors where it can be properly tagged, identified, and photographed.
    C. have the building owner walk through the building to pick up valuables.
    D. leave suspected evidence where it is found.

_____ **48.** Initial recognition and preservation of evidence is the responsibility of the:
  A. fire marshal.
  B. fire inspector.
  C. fire investigator.
  D. fire fighter.

_____ **49.** When cribbing is used for vehicle stabilization,
  a _____ formation is most often
  used to support and stabilize vehicles.
  A. wedging
  B. pyramid
  C. box
  D. straight stack

_____ **50.** In patient removal, the term used to dress and bandage, splint fractures, and immobilize the patient's body to reduce possibility of further injury is called:
  A. advanced life support care.
  B. packaging.
  C. full body immobilization.
  D. all clear.

_____ **51.** When using a spring-loaded center punch to open a window, the fire fighter should press the center punch at the:
  A. center of the window.
  B. lower corner of the window.
  C. upper corner of the window.
  D. at the top of the window.

_____ **52.** _____ glass is hardened glass designed to shatter into small pieces.
  A. Tempered
  B. Plastic-coated
  C. Laminated
  D. Shock-resistant

_____ **53.** You are operating as a member of a rescue company at the scene of an auto extrication. The scene, vehicle, and the patient(s) are stabilized, all hazards are controlled, and access and disentanglement have been accomplished. The next step in the extrication process should be:
  A. termination and post-incident analysis.
  B. patient packaging and patient removal.
  C. transporting patient(s) to appropriate facilities.
  D. initiating extrication operations on another vehicle, if needed.

_____ **54.** Which of the following statements regarding air bags is **<u>incorrect</u>**?
  A. Bags must be on or against a solid base.
  B. You must crib as you lift.
  C. Do not stack more than two bags.
  D. When stacking two bags of different sizes the larger bag goes on top.

_____ **55.** The reason that many departments have a policy of placing apparatus at an angle to the crash site at a vehicle accident is:

    A. because this position presents the widest profile of flashing lights to oncoming traffic.

    B. because then the pump operator can see through the walkway of a mid-mount apparatus and watch the scene as well as oncoming traffic.

    C. so that if it is struck from behind by an oncoming vehicle, it will be pushed off to the side.

    D. so that it will present the largest possible barrier profile to oncoming traffic.

_____ **56.** What absolutely must be used in conjunction with rescue lift airbags?

    A. Cribbing

    B. Hose tape

    C. Blocking

    D. Designator ribbon

_____ **57.** If a windshield must be removed to gain access, what tool is appropriate?

    A. An ax

    B. A spring-loaded center punch

    C. A screwdriver

    D. An Allen wrench (6 mm)

_____ **58.** When using pneumatic air bags to perform a lift, one should never stack **more** than _____ bags.

    A. two

    B. three

    C. four

    D. five

_____ **59.** The process of erecting materials such as wood panels, timber, or jacks to strengthen a wall or prevent further collapse is known as:

    A. shoring.

    B. cribbing.

    C. packing.

    D. supporting.

_____ **60.** The **most** **common** hazard encountered at a confined space incident is:

    A. an oxygen deficient atmosphere.

    B. engulfment.

    C. collapse of structure.

    D. entrapment.

_____ **61.** What is the step in the technical rescue incident process called in which victims are freed from entrapment?

    A. Access

    B. Recovery

    C. Release

    D. Disentanglement

_____ **62.** Concrete has excellent _____ strength when it cures.
   A. shear
   B. compressive
   C. torsional
   D. tensile

_____ **63.** What is the simplest and **most effective** method of achieving the fire service goal of the preservation of life and property?
   A. Prevention
   B. Improved technology
   C. More fire fighters
   D. More fire stations

_____ **64.** Building construction is an important factor to be identified in a preincident survey. While conducting a survey on a property, you observe the structure members are of noncombustible materials but may not have fire resistive protection. How would you identify this construction type on the survey form?
   A. Type I
   B. Type II
   C. Type III
   D. Type IV

_____ **65.** A facility in which there is a great potential likelihood of life or property loss from a fire is called a/an _____ hazard.
   A. special
   B. assembly
   C. target
   D. industrial

_____ **66.** Some fire departments mention other types of hazards than just those specifically fire-related to occupants during residential fire safety surveys. What is one of these other types of hazards?
   A. Carbon monoxide
   B. Radon testing
   C. Driving under the influence
   D. Hurricane roof tie down

_____ **67.** Which of the following is a factor in deciding to preplan a structure or area?
   A. Type of hazards expected
   B. Complexity of firefighting operations
   C. Nature of activities conducted at the occupancy
   D. All of the above.

_____ **68.** The primary purpose of conducting a residential fire safety survey is to:
   A. pre-plan residential areas.
   B. look for illegal activities.
   C. reduce loss of life and property.
   D. impress the city council.

_____ **69.** The goal of fire prevention activities is to accomplish all of the following <u>except</u>:
   A. reducing the dangers to fire fighters.
   B. receiving appropriate media coverage of emergencies.
   C. reducing the risk of public safety.
   D. reducing the amount of property damage.

_____ **70.** One topic a fire fighter may be asked to present during a fire safety presentation to an external group is:
   A. fire stream applications.
   B. home safety practices.
   C. fire extinguisher maintenance.
   D. rescue practices.

_____ **71.** Heat detectors:
   A. cannot be used as part of a suppression system.
   B. are slow to activate.
   C. are expensive to install and operate.
   D. are responsible for most false alarms.

_____ **72.** Smoke detectors work primarily on the principles of photoelectricity and:
   A. rate of rise.
   B. fixed temperature.
   C. ionization.
   D. laser beam.

_____ **73.** The purpose of the fire department connection to a sprinkler system is to:
   A. supplement the water supply while maintaining operational pressure.
   B. provide water, since most systems are dependent on the fire department for water supply.
   C. boost the water to upper stories, since most water pressure is <u>not</u> sufficient to supply water above the sixth floor.
   D. add water pressure to the system because normal water distribution is inadequate when less than three heads are activated.

_____ **74.** The control valve for a sprinkler system may be located at the system or outside the building. This valve should always be a/an _____ valve.
   A. check
   B. indicating-type
   C. inspector's test
   D. quarter-turn

_____ **75.** Which of the following sprinkler system components is used to limit water flow to one direction?
   A. OS&Y valve
   B. Check valve
   C. Control valve
   D. Butterfly valve

_____ **76.** Direct pumping water systems are those in which water:
  A. moves directly into the distribution system by gravity flow.
  B. is supplied directly into the distribution system from elevated storage tanks.
  C. is pumped directly into the distribution system with no elevated storage.
  D. is pumped directly through the distribution system back into the main water supply.

_____ **77.** Which of the following violates the principle of a loop feed hydrant?
  A. Primary feeders
  B. Secondary feeders
  C. Interconnecting distributors
  D. Dead-end water mains

_____ **78.** Which of the following **is** **not** one of the component parts of a dry-barrel fire hydrant?
  A. Operating stem
  B. Stem nut
  C. Post-indicator gate
  D. Drain hole

_____ **79.** The following illustration depicts a _____ hydrant.
  A. dry-barrel
  B. wet-barrel
  C. drafting
  D. dry

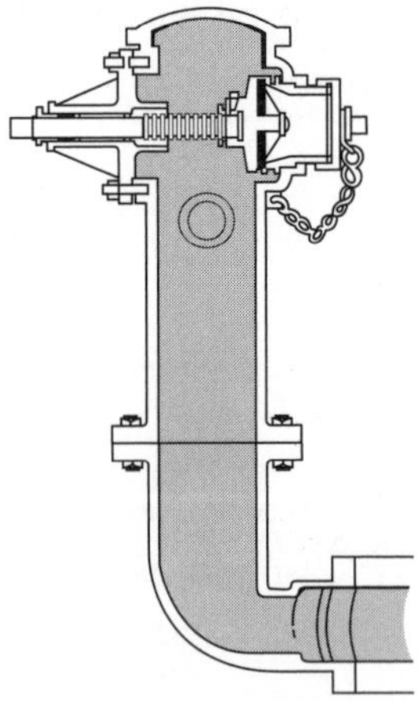

_____ **80.** A fire hydrant that receives water from only one direction is called a _____ hydrant.
  A. one-way
  B. steamer
  C. circulating-feed
  D. dead-end

_____ **81.** The smaller internal grid arrangement of a water distribution system that feeds hydrants, as well as the domestic and commercial requirements, **best** describes:
  A. primary feeders.
  B. secondary feeders.
  C. distributors.
  D. grid network.

_____ **82.** Large pipes that carry large quantities of water to various points along the water supply system for distribution to smaller mains **best** defines:
  A. primary feeders.
  B. secondary feeders.
  C. distributors.
  D. grid network.

_____ **83.** The following illustration depicts a _____ hydrant.
  A. dry-barrel
  B. wet-barrel
  C. drafting
  D. dry

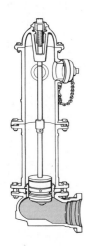

_____ **84.** A network of intermediate-sized pipe that reinforces the overall grid system by forming loops that interlock primary feeders **best** defines:
  A. primary loop.
  B. secondary feeders.
  C. distributors.
  D. grid network.

_____ **85.** Which statement regarding residential sprinklers is **incorrect**?
  A. They are smaller versions of wet- or dry-pipe sprinkler systems.
  B. The water supply is combined with the domestic system.
  C. The use of plastic pipe is allowed.
  D. They are designed to control the level of fire involvement such that residents can escape.

_____ **86.** A post indicator valve (PIV) is:
  A. a device to speed the operation of the dry-pipe valve by detecting the decrease in air pressure.
  B. designed to control the head pressure at the outlet of a standpipe system to prevent excessive nozzle pressures in hose lines.
  C. a control valve that is mounted on a post case with a small window reading either "open" or "shut."
  D. a control valve that is mounted on a wall in a metal case with a small window reading either "open" or "shut."

_____ **87.** Dry-pipe systems are used in all of the following incidents **except**:
  A. in buildings that refrigerate or freeze materials.
  B. in unheated buildings.
  C. outdoor applications where freezing temperatures occur.
  D. where rapid activation is required.

_____ **88.** **Directions**: Read the following statements regarding fire hydrants and select your answer from choices A–D.
  1. Mechanical damage to fire hydrants can be caused by many things Including nature, vandals, accidents, and improper actions by members of the fire department.
  2. Rust and corrosion of hydrants should be noted for referral to the water department during hydrant inspection by fire fighters.
  3. Sedimentation and encrustation can restrict or completely obstruct flow from hydrants.

  A. Statement 1 is true; statements 2 and 3 are false.
  B. Statements 1 and 2 are true; statement 3 is false.
  C. Statement 1 is false; statements 2 and 3 are true.
  D. All three statements are true.

_____ **89.** The simplest sprinkler system in design and operation is the _____ system.
  A. wet-pipe
  B. deluge
  C. dry-pipe
  D. preaction

_____ **90.** Under normal circumstances, the air pressure gauge on a dry-pipe sprinkler system will read _____ the water pressure gauge.
  A. the same as
  B. higher than
  C. lower than
  D. almost double

_____ **91.** A sprinkler head with a temperature rating of 135°–170°F would be:
  A. orange.
  B. red.
  C. white.
  D. either uncolored or black.

_____ **92.** Which of the following <u>is</u> <u>not</u> considered to be an alarm initiating device?
  A. Water detector
  B. Visible products-of-combustion detector
  C. Flame detector
  D. Invisible products-of-combustion detector

_____ **93.** When filling a hose line with water for testing, a pump pressure of approximately _____ psi is maintained.
  A. 25 to 30
  B. 45 to 50
  C. 80 to 90
  D. 250

_____ **94.** When conducting a hose service test, after charging the hose line with water:
  A. ensure all air has been discharged from the line.
  B. have an adequate number of fire fighters controlling each line.
  C. tighten each coupling as tightly as possible.
  D. walk each line to determine whether there are any air leaks.

_____ **95.** After the proper test pressure has been reached during an annual service test on fire hose, pressure should be maintained for a period of:
  A. thirty seconds.
  B. one to three minutes.
  C. three to five minutes.
  D. twenty minutes.

_____ **96.** The <u>**maximum**</u> length of time that fire hose should be used without a service test is:
  A. six months.
  B. one year.
  C. three years.
  D. five years.

_____ **97.** When laying out fire hose to be service tested, test lengths should be:
  A. no more than 150 feet.
  B. no more than 300 feet.
  C. no more than 500 feet.
  D. no more than 200 feet.

_____ **98.** To determine whether there is any slippage of couplings when testing hose:
  A. measure and mark the exact length of each coupling.
  B. mark the hose by each coupling using a pencil/marker.
  C. check to see whether couplings remain tight.
  D. use a special torque wrench set to 60 psi.

_____ **99.** When testing fire hose, fill each hose with water and, during the filling process, make sure that each nozzle:
  A. receives the same gpm.
  B. is closed and strapped in place.
  C. is held by a fire fighter when discharging water.
  D. is open until all air is discharged.

_____**100.** The type of hose testing conducted by the fire department is called:
   A. acceptance.
   B. pressure.
   C. service.
   D. fireground.

Did you score higher than 80 percent on Examination II-3? Circle Yes or No in ink.

Feedback Step

Now, what do we do with your "yes" and "no" answers given throughout the SAEP process? First, return to any response that has "no" circled. Go back to the highlighted answers for those examination items missed. Read and study the paragraph preceding the location of the answer, as well as the paragraph following the paragraph where the answer is located. This will expand your knowledge base for the missed question, put it in a broader perspective, and improve associative learning. Remember, you are trying to develop mastery of the required knowledge. Scoring 80 percent on an examination is good, but it is not mastery performance. To be at the top of your group, you must score much higher than 80 percent on your training, promotion, or certification examination.

Phases III and IV focus on getting you ready for the examination process by recommending activities that have a positive impact on the emotional and physical aspects of examination preparation. By evaluating your own progress through SAEP, you have determined that you possess a high level of knowledge. Taking an examination for training, promotion, or certification is a competitive event. Just as in sports, total preparation is vitally important. Now you need to get all the elements of good preparation in place so that your next examination experience will be your best ever. Phase III is next! First, however, review the Summary of Key Rules for Taking an Examination and Summary of Helpful Hints.

Summary of Key Rules for Taking an Examination

Rule 1—Examination preparation is not easy. Preparation is 95% perspiration and 5% inspiration.

Rule 2—Follow the steps very carefully. Do not try to reinvent or shortcut the system. It really works just as it was designed to!

Rule 3—Mark with an "X" any examination items for which you guessed the answer. For maximum return on effort, you should also research any answer that you guessed even if you guessed correctly. Find the correct answer, highlight it, and then read the entire paragraph that contains the answer. Be honest and mark all questions on which you guessed. Some examinations have a correction for guessing built into the scoring process. The correction for guessing can reduce your final examination score. If you are guessing, you are not mastering the material.

Rule 4—Read questions twice if you have any misunderstanding, especially if the question contains complex directions or activities.

Rule 5—If you want someone to perform effectively and efficiently on the job, the training and testing program must be aligned to achieve this result.

Rule 6—When preparing examination items for job-specific requirements, the writer must be a subject matter expert with current experience at the level that the technical information is applied.

Rule 7—Good luck = good preparation.

Summary of Helpful Hints

Helpful Hint—Most of the time your first impression is the best. More than 41% of changed answers during our SAEP field test were changed from a right answer to a wrong answer. Another 33% were changed from a wrong answer to another wrong answer. Only 26% of changed answers were changed from wrong to right. In fact, three participants did not make a perfect score of 100% because they changed one right answer to a wrong one! Think twice before you change your answer. The odds are not in your favor.

<u>Helpful Hint</u>—Researching correct answers is one of the most important activities in SAEP. Locate the correct answer for all missed examination items. Highlight the correct answer. Then read the entire paragraph containing the answer. This will put the answer in context for you and provide important learning by association.

<u>Helpful Hint</u>—Proceed through all missed examination items using the same technique. Reading the entire paragraph improves retention of the information and helps you develop an association with the material and learn the correct answers. This step may sound simple. A major finding during the development and field testing of SAEP was that you learn from your mistakes.

<u>Helpful Hint</u>—Follow each step carefully to realize the best return on effort. Would you consider investing your money in a venture without some chance of earning a return on that investment? Examination preparation is no different. You are investing time and expecting a significant return for that time. If, indeed, time is money, then you are investing money and are due a return on that investment. Doing things right and doing the right things in examination preparation will ensure the maximum return on effort.

<u>Helpful Hint</u>—Try to determine why you selected the wrong answer. Usually something influenced your selection. Focus on the difference between your wrong answer and the correct answer. Carefully read and study the entire paragraph containing the correct answer. Highlight the answer.

<u>Helpful Hint</u>—Studying the correct answers for missed items is a critical step in achieving your desired return on effort! The focus of attention is broadened and new knowledge is often gained by expanding association and contextual learning. During PTS's research and field test, self-study during this step of SAEP resulted in gains of 17 points between the first examination administered and the third examination. A gain of 17 points can move you from the lower middle to the top of the list of persons taking a training, promotion, or certification examination. That is a competitive edge and a prime example of return on effort in action. Remember: Maximum effort = maximum results!

PHASE III

How Examination Developers Think—Getting Inside Their Heads

Now that you've finished the examination practice, this phase will assist you in understanding and applying examination-taking skills. Developing your knowledge of how examination professionals think and prepare examinations is not cheating. Most serious examination takers have spent many hours reviewing various examinations to gain an insight into the technology used to develop them. It is a demanding technology when used properly. You probably already know this if you have prepared examination items and administered them in your fire department.

Phase III will not cover all the ways and means of examination-item writing. Examination-item writers use far too many techniques to cover adequately in this book. Instead, the focus here is on key techniques that will help you achieve a better score on your examination.

How are examination items derived?

Professional examination-item writers use three basic techniques to derive examination items from text or technical reference materials: verbatim, deduction, and inference.

The most common technique is to take examination items verbatim from materials in the reference list. This technique doesn't work well for mastering information, however. The verbatim form of testing encourages rote learning—that is, simply memorizing the material. The results of this type of learning are not long-lasting, nor are they appropriate for learning and retaining the critical knowledge that you must have for on-the-job performance. Consequently, SAEP doesn't create the majority of examination questions covering NFPA 1001 using the verbatim technique.

Professional examination-item writers tend to use verbatim testing at the most basic level of job classifications. A fire fighter, for instance, is expected to learn many basic facts. At this level, verbatim examination items can be justified.

In the higher ranks of the Fire and Emergency Medical Service, other methods are more beneficial and productive for mastering higher cognitive knowledge and skills. At the higher cognitive levels of an occupation, such as fire officer, examination development will therefore rely on other means. The most important technique at the higher cognitive levels is using deduction as the basis for examination items. This technique requires logic and analytical skills and often requires the examination taker to read materials several times to answer the examination item. It is not, then, a matter of simply repeating the information that results in a verbatim answer.

At the Fire Fighter I level, most activities are carefully supervised by a more experienced fire fighter or Company Officer. At this level, the responder is expected to closely follow commands and is encouraged not to use deductive reasoning that can lead to "freelance" responder tactics. As one progresses to a higher level job and gains experience, deductive reasoning and inference skills are developed and applied. Most of these skills are related to personal safety and the safety of those on the scene. Most sizeups and strategies are developed and passed from the officers on the scene to the first responders.

Rule 5

If you want someone to perform effectively and efficiently on the job, the training and testing program must be aligned to achieve this result.

Rule #5 is of paramount importance for fire fighters. Effective and efficient fire fighters are able to receive fireground commands, follow instructions, and perform their tasks as safely and as rapidly as they can. There are limited opportunities for fire fighters to do much else, because they serve as the first line of action at the emergency scene.

Consider the following example of deductive reasoning: An incident call is received from the telecommunicator stating that an infant has a high temperature and is convulsing. Just this amount of information should cause the fire fighter to immediately plan the response, conduct sizeup activities, and review infant care procedures en route. Some of these deductive responses will have you focus on the infant's age, past medical history, location, access, and many other possible factors. If you have an EMT or paramedic background, a list of several items could be deduced that would expedite an efficient and effective response to the incident.

You can probably think of many firefighting tasks and circumstances that rely on deductive reasoning. The more experience you gain on the fireground as a fire fighter, the more often you will be called upon to practice deductive reasoning and inference from emergency data, and the more efficient and effective you will become, whether the situation involves ventilating a roof or attending to the emergency needs of an infant.

Legendary football coach Vince Lombardi was once asked about the precision performance of his offensive and defensive teams. It was suggested that Lombardi must spend a great deal of time on the practice field to achieve those results. Lombardi responded, "Practice doesn't make perfect; only perfect practice makes perfect." This is exactly what is required to be an outstanding examination taker. Most people don't perfectly practice examination-taking skills.

A third technique used by professional examination-item writers is to rely on inference or implied answers to develop examination items. Inference requires contrasting, comparing, analyzing, evaluating, and other high-level cognitive skills. Tables, charts, graphs, and other instruments for presenting data provide excellent means for deriving inference-based examination items. Implied answers are based on logic. They rely on your ability to use logical processes or series of facts to arrive at a plausible answer.

For example, recent data gathered by the NFPA stated that heart attacks remain the leading cause of death for fire service personnel. Other NFPA-supplied data indicated that strains and sprains are the leading cause of injuries on the job. Several inferences can be made from these relatively simple statements. A safety officer can infer the results of the NFPA study to his or her own personnel and use the information as a trigger for checking on personnel, conducting surveys, reviewing accident records, and comparing the study results with actual experience. Is that particular fire department doing better or worse in terms of these important health issues? Are the Fire and Emergency Medical Service personnel getting the right exercise? Are they diligent in keeping the station and fireground free from the activities that may lead to strains and sprains? The basic inference here is that any particular fire department may be similar or different in some ways from the generalized data.

Sometimes it may be difficult to find an answer to an examination item because it is measuring your ability to make deductions and draw inferences from the technical materials.

How are examination items written and validated?

Once the pertinent information is identified and the technique for writing an examination item selected, the professional examination-item writer will prepare a draft. The draft examination item is then referenced to specific technical information, such as a textbook, manufacturer's manual, or other related technical information. If the information is derived from a job-based requirement, then it should also be validated by job incumbents (i.e., those who are actually performing in the occupation at the specific level of the required knowledge).

--- **Rule 6** ---

When preparing examination items for job-specific requirements, the writer must be a subject matter expert with current experience at the level where the technical information is applied.

Rule #6 ensures that the examination item has a basic level of job content validity. The final level of job content validity is determined by using committees or surveys of job incumbents who certify the information to be current and required on the job. The information must be in a category of "need to know" or "must know" to be considered job relevant. The technical information must be accurate. Because subject matter experts do need basic training in examination-item writing, it is recommended that a professional in examination technology be part of the review process so that basic rules and guidelines of the industry are followed.

Finally, the examination items must be field tested. Once this testing is complete, statistical and analytical tools are available to help revise and improve the examination items. These techniques and tools go well beyond the scope of this *Exam Prep* book. Professionals are available to conduct these data analyses, and their services should be used.

Good Practices in Examination Item and Examination Development

The most reliable examinations are objective. That is, each question has only one answer that is accepted by members of the occupation. This objective quality permits fair and equitable examinations. The most popular types of objective examination items are multiple choice, true/false, matching, and completion (fill in the blanks).

Valid and reliable job-relevant examinations for the Fire and Emergency Medical Service industry must satisfy 10 rules:

1. They do not contain trick questions.
2. They are short and easy to read, using language and terms appropriate to the target examination population.
3. They are supported with technical references, validation information, and data on their difficulty, discrimination, and other item analysis statistics.
4. They are formatted to meet recognized testing standards and examples.
5. They focus on the "need to know" and "must know" aspects of the job.
6. They are fair and objective.
7. They are not based on obscure and trivial knowledge and skills.
8. They can be easily defended in terms of job-content requirements.
9. They meet national and other professional job qualification standards.
10. They demonstrate their usefulness as part of a comprehensive testing program, including written, oral, and performance examination items.

The primary challenges of job-relevant examinations relate to their currency and validity. Careful recording of data, technical reference sources, and the examination writer's qualifications are important. Examinations that affect someone's ability to be promoted, certified, or licensed, as well as to complete training that leads to a job, have exacting requirements both in published documents and in the laws of the land.

Three Common Myths of Examination Construction

1. **Myth:** If in doubt about the answer for a multiple-choice examination item, select the longest answer.

   Reality: Professional examination-item writers use short answers as correct ones at an equal or higher percentage than longer answers. Remember, there are usually choices A–D. That leaves three other possibilities for the correct answer other than the longest one. Statistically speaking, the longest answer is less likely to be correct.

2. **Myth:** If in doubt about the answer for a multiple-choice examination item, select "C".

   Reality: Computer technology and examination-item banking permit multiple versions of examinations to be developed simultaneously. This is typically achieved by moving the correct answer to different locations (for example, version 1 will have the correct answer in the "C" position, version 2 will have it in the "D" position, and so forth).

3. **Myth:** Watch for errors in singular examination-item stems with plural choices in the A–D answers, or vice versa.

   Reality: Most computer-based programs have spelling and grammar checking utilities. If this mistake occurs, an editing error is the probable cause and usually has nothing to do with detecting the right answer.

Some Things That Work

1. Two to three days before your examination, review the examination items you missed in SAEP. Read those highlighted answers and the entire paragraph one more time.
2. During the examination, carefully read each examination item twice. Once you have selected your answer, read the examination item and answer together. This technique can prompt you to recall information that you studied during your examination preparation activities.
3. Apply what you learned in SAEP. Eliminate as many distracters as possible to improve your chance of answering the question correctly.
4. Pace yourself. Know how much time you have to take the examination. If an examination item is requiring too much time, write its number down and continue with the next examination item. Often, a later examination item will trigger your memory and make the earlier examination item seem easier to answer. (For a time pacing strategy, see the Examination Pacing Table at the end of Phase IV.)
5. Don't panic if you don't know some examination items. Leave them to answer later. The most important thing is to finish the examination, because there may be several examination items at the end of the examination that you do know.

6. As time runs out for taking the examination, do not panic. Concentrate on answering those difficult examination items that you skipped.

7. Double-check your answer sheets to make sure you have not accidentally left an answer blank.

8. Once you complete the examination, return to the difficult examination items. Often, while taking an examination, other examination items will cause you to remember or associate those answers with the difficult examination-item answers. The longer the examination, the more likely you will be to gather the information needed to answer more difficult examination items.

There are many other helpful hints that can be used to improve your examination-taking skills. If you want to research the materials on how to take examinations and raise your final score, visit your local library, a bookstore, or the Web for additional resources. The main reason we developed SAEP is to provide practice and help you develop examination-taking skills that you can use throughout your life.

PHASE IV

The Basics of Mental and Physical Preparation

Mental Preparation—I Can Get My Head Ready!

The two most common mental blocks to examination taking are examination anxiety and fear of failure. In the Fire and Emergency Medical Service, these feelings can create significant performance barriers. Overcoming severe conditions may require some professional psychological assistance, which is beyond the scope of this *Exam Prep* book.

The root cause of examination anxiety and fear of failure is often lack of self-confidence. SAEP was designed to help improve your self-confidence by providing evidence of your mastery of the material on the examination. Look at your scores as you progress through Phase I or II. Review your Personal Progress Plotter; it will help you gain confidence in your knowledge of NFPA 1001. Look at your Personal Progress Plotter the day before your scheduled examination and experience renewed confidence.

Let's examine the meaning of anxiety. Knowing what it is will help you deal with it at examination time. According to *Webster's Dictionary*, anxiety is "uneasiness and distress about future uncertainties." Many of us have real anxiety about taking examinations, and it is a natural response for some, often prefaced by questions like these: Am I ready for this? Do I have a good idea of what will be on the examination? Will I make the lowest score? Will John Doe score higher than me?

These questions and concerns are normal. Remember that hundreds of people have gone through SAEP and achieved an average gain of 17 points in their scores. The preparation process will help you maintain your self-confidence. Once again, review the evidence in your Personal Progress Plotter to see what you have accomplished.

Fear, according to *Webster's Dictionary*, is "alarm and agitation caused by the expectation or realization of danger." It is a normal reaction to examinations. To deal with it, first analyze the degree of fear you may be experiencing several days before the examination date. Then focus on the positive experiences you had as you finished SAEP. Putting your fear in perspective by using positives to eliminate or minimize it is a very important examination-taking skill. The more you focus on your positive accomplishments in mastering the materials, the less fear you will experience.

If your fear and anxiety persist even after you take steps to build your confidence, you may want to get some professional assistance. Do it now! Don't wait until the week before the examination. There may be real issues that a professional can help you deal with to overcome these feelings. Hypnosis and other forms of treatment have been found to be very helpful. Consult with an expert in this area.

Physical Preparation—Am I Really Ready?

Physical preparation is the element that is probably most ignored in examination preparation. In the Fire and Emergency Medical Service, examinations are often given at locations away from home. If this is the case, you need to be especially careful of key physical concerns. More will be said about that later.

In general, following these helpful hints will help you concentrate, enhance your examination performance, and add points to your score.

1. Do not "cram" for the examination. This factor was found to be first in importance during PTS's field test of SAEP. Cramming results in examination anxiety, adds to confusion, and tends to lessen the effectiveness of the examination-taking skills you already possess. Avoid cramming!

2. Get a normal night's rest. It may even be wise to take a day off before the examination to rest. Do not schedule an all-night shift right before your examination.

3. Avoid taking excessive stimulants or medications that inhibit your thinking processes. Eat at least three well-balanced meals before the day of the examination. It is a good practice to carry a balanced energy bar (not candy) and a bottle of water into the examination area. Examination anxiety and fear can cause a dry mouth, which can lead to further aggravation. Nibbling on the energy bar also has a settling effect and supplies some "brain food."

4. If the examination is taking place at an out-of-town location, do the following:
   • Avoid a "night out with friends." Lack of rest, partying, and fatigue are major examination performance killers.
   • Check your room carefully. Eliminate things that may aggravate you, interfere with your rest, or cause any discomfort. If the mattress isn't good, the pillows are horrible, or the room has an unpleasant odor, change rooms or even hotels.
   • Wake up in plenty of time to take a relaxing shower or soaking bath. Don't put yourself in a "rush" mode. Things should be carefully planned so that you arrive at the examination site ahead of time, calm, and collected.

5. Listen to the examination proctor. The proctor usually has rules that you must follow. Important instructions and directions are usually given. Ask clarifying questions immediately and listen to the responses to questions raised by the other examination takers. Most examination environments are carefully controlled and may not permit questions you raise that are covered in the proctor's comments or deal with the technical content in the examination itself. Be attentive, focus, and succeed.

6. Remain calm and breathe. Pace yourself. Apply your examination-taking skills learned during SAEP.

7. Remember the analogy of an examination as a competitive event. If you want to gain a competitive edge, carefully follow all phases of SAEP. This process has yielded outstanding results in the past and will do so for you.

Time Management During Examinations

The following table will help you pace yourself during an examination. You should become familiar with the table and be able to construct your own when you are in the examination room and getting ready to start the examination process. This effort will take a few minutes, but it will make a tremendous contribution to your time management during the examination.

Here is how the table works. First you divide the examination time into 6 equal parts. For example, if you have 3.5 hours (210 minutes) for the examination, then each of the 6 time parts contains 35 minutes (210 ÷ 6 = 35 minutes). Now divide the number of examination items by 5. For example, if the examination has 150 examination items, 150 ÷ 5 = 30. Now, with the math done, we can set up a table that tells you approximately how many examination items you should answer in 35 minutes (the equal time divisions). You should be on or near examination item 30 at the end of the first 35 minutes and so forth. Notice that we divided the number of examination items by 5 and the time available by 6. The extra time block of 35 minutes is used to double-check your answer sheet, focus on difficult questions, and calm your nerves. This technique will work wonders for your stress level, and yes, it will improve your examination score.

Examination Pacing Table (150 and 100 Examination Items)

Time for Examination	Minutes for 6 Equal Time Parts	Number of Examination Items	Examination Items Per Time Part	Time for Examination Review
210 Minutes (3.5 Hours)	35 Minutes	150	30 (# of examination items to be answered)	35 Minutes (Chilling and Double-checking Examination)
150 Minutes (2.5 Hours)	25 Minutes	100	20 (# of examination items to be answered)	(Chilling and Double-checking Examination)

Using the Examination Pacing Table can be simplified by using the time/examination item variables as either may change in the real examination environment. For instance, if time is changed, adjust the ratio of time to answer the examination items in each of the five time blocks. If examination item numbers increase or decrease, adjust the number of examination items to be answered in the time blocks.

Some precautions when using this time management strategy:

1. Do not panic if you run a few minutes behind in each time block. It is a time management strategy and should not stress you while using it. Most people tend to pick up their pace as they move into the examination.

2. During the examination carefully mark or note examination items that you need to return to during your review time block. This will help you expedite your examination completion check.

3. Do not be afraid to ask for more time to complete your examination. In most cases, the time limit is flexible or should be.

4. Last, but not least, double check your answer sheet to make sure you didn't leave blank responses and that you didn't double mark answers. Double markings are most often counted as wrong answers. Make sure any erasers are made cleanly. Caution: Make sure when you change your answer that you really want to do that. Odds are not in your favor unless something on the examination really influenced the change.

APPENDIX A

Examination I-1 Answer Key

Directions

Follow these steps carefully for completing the feedback part of SAEP:

1. After calculating your score, look up the answers for the examination items you missed as well as those on which you guessed, even if you guessed correctly. If you are guessing, it means the answer is not perfectly clear. In this process, we are committed to making you as knowledgeable as possible.

2. Enter the number of missed and guessed examination items in the blanks on your Personal Progress Plotter.

3. Highlight the answer in the reference materials. Read the paragraph preceding and the paragraph following the one in which the correct answer is located. Enter the paragraph number and page number next to the guessed or missed examination item on your examination. Count any part of a paragraph at the beginning of the page as one paragraph until you reach the paragraph containing your highlighted answer. This step will help you locate and review your missed and guessed examination items later in the process. This step is essential to learning the material in context and by association. These learning techniques (context/association) are the very backbone of the SAEP approach.

4. Once you have completed the feedback part, you may proceed to the next examination.

1. Reference: NFPA 1001, 5.1.1
   Delmar, *Firefighter's Handbook*, 3rd Edition, 1st Printing, pages 26 and 28.
   IFSTA, *Essentials of Fire Fighting and Fire Department Operations*, 5th Edition, 1st Printing, page 9.
   Jones and Bartlett, NFPA, *Fundamentals of Fire Fighter Skills*, 2nd Edition, 1st Printing, page 2.
   Answer: A

2. Reference: NFPA 1001, 5.1.1, 3.3.1, 5.3.9, and 5.3.9(A)
   Delmar, *Firefighter's Handbook*, 3rd Edition, 1st Printing, page 29.
   IFSTA, *Essentials of Fire Fighting and Fire Department Operations*, 5th Edition, 1st Printing, page 21.
   Jones and Bartlett, NFPA, *Fundamentals of Fire Fighter Skills*, 2nd Edition, 1st Printing, page 6.
   Answer: A

3. Reference: NFPA 1001, 5.1.1, 1.3.1, and 1.2

Delmar, *Firefighter's Handbook*, 3rd Edition, 1st Printing, page 109.

IFSTA, *Essentials of Fire Fighting and Fire Department Operations*, 5th Edition, 1st Printing, page 51.

Jones and Bartlett, NFPA, *Fundamentals of Fire Fighter Skills*, 2nd Edition, 1st Printing, page 26.

Answer: D

4. Reference: NFPA 1001, 5.1.1

Delmar, *Firefighter's Handbook*, 3rd Edition, 1st Printing, page 42.

IFSTA, *Essentials of Fire Fighting and Fire Department Operations*, 5th Edition, 1st Printing, page 36.

Jones and Bartlett, NFPA, *Fundamentals of Fire Fighter Skills*, 2nd Edition, 1st Printing, page 111.

Answer: D

5. Reference: NFPA 1001, 5.1.1, 4.2

Delmar, *Firefighter's Handbook*, 3rd Edition, 1st Printing, page 831.

IFSTA, *Essentials of Fire Fighting and Fire Department Operations*, 5th Edition, 1st Printing, page 168.

Jones and Bartlett, NFPA, *Fundamentals of Fire Fighter Skills*, 2nd Edition, 1st Printing, page 586.

Answer: D

6. Reference: NFPA 1001, 5.1.1, 5.3.1, 5.3.1(A), 5.3.9, 5.3.9(A), 5.3.10, 5.3.10(A), 5.3.11, 5.3.11(A), 5.3.12, 5.3.12(A), 5.3.13 and 5.3.13(A)

Delmar, *Firefighter's Handbook*, 3rd Edition, 1st Printing, page 705.

IFSTA, *Essentials of Fire Fighting and Fire Department Operations*, 5th Edition, 1st Printing, page 793.

Jones and Bartlett, NFPA, *Fundamentals of Fire Fighter Skills*, 2nd Edition, 1st Printing, page 588.

Answer: A

7. Reference: NFPA 1001, 5.1.1, 5.3.4, and 5.3.4(A)

Delmar, *Firefighter's Handbook*, 3rd Edition, 1st Printing, page 771.

IFSTA, *Essentials of Fire Fighting and Fire Department Operations*, 5th Edition, 1st Printing, page 396.

Jones and Bartlett, NFPA, *Fundamentals of Fire Fighter Skills*, 2nd Edition, 1st Printing, page 650 and 512.

Answer: D

8. Reference: NFPA 1001, 5.1.1, and 5.1.2

Delmar, *Firefighter's Handbook*, 3rd Edition, 1st Printing, page 488.

IFSTA, *Essentials of Fire Fighting and Fire Department Operations*, 5th Edition, 1st Printing, pages 277 and 292, Skill Sheet 7-1–7.

Jones and Bartlett, NFPA, *Fundamentals of Fire Fighter Skills*, 2nd Edition, 1st Printing, page 257.

Answer: D

9. Reference: NFPA 1001, 5.1.1 and 5.1.2

Delmar, *Firefighter's Handbook*, 3rd Edition, 1st Printing, page 476.

IFSTA, *Essentials of Fire Fighting and Fire Department Operations*, 5th Edition, 1st Printing, page 267.

Jones and Bartlett, NFPA, *Fundamentals of Fire Fighter Skills*, 2nd Edition, 1st Printing, page 242.

Answer: C

10. Reference: NFPA 1001, 5.1.1 and 5.1.2

Delmar, *Firefighter's Handbook*, 3rd Edition, 1st Printing, page 483.

IFSTA, *Essentials of Fire Fighting and Fire Department Operations*, 5th Edition, 1st Printing, page 290.

Jones and Bartlett, NFPA, *Fundamentals of Fire Fighter Skills*, 2nd Edition, 1st Printing, page 253.

Answer: C

11. Reference: NFPA 1001, 5.1.1, and 5.1.2

Delmar, *Firefighter's Handbook*, 3rd Edition, 1st Printing, page 505.

IFSTA, *Essentials of Fire Fighting and Fire Department Operations*, 5th Edition, 1st Printing, pages 281 and 295.

Jones and Bartlett, NFPA, *Fundamentals of Fire Fighter Skills*, 2nd Edition, 1st Printing, page 264.

Answer: D

12. Reference: NFPA 1001, 5.1.1

Delmar, *Firefighter's Handbook*, 1st Printing, page 1104.

IFSTA, *Essentials of Fire Fighting and Fire Department Operations*, 5th Edition, 1st Printing, page 19.

Jones and Bartlett, NFPA, *Fundamentals of Fire Fighter Skills*, 2nd Edition, 1st Printing, page 8.

Answer: A

13. Reference: NFPA 1001, 5.1.1

Delmar, *Firefighter's Handbook*, 3rd Edition, 1st Printing, page 46.

IFSTA, *Essentials of Fire Fighting and Fire Department Operations*, 5th Edition, 1st Printing, page 39.

Jones and Bartlett, NFPA, *Fundamentals of Fire Fighter Skills*, 2nd Edition, 1st Printing, page 113.

Answer: A

14. Reference: NFPA 1001, 5.1.1

Delmar, *Firefighter's Handbook*, 3rd Edition, 1st Printing, pages 689, 908, and 1117.

IFSTA, *Essentials of Fire Fighting and Fire Department Operations*, 5th Edition, 1st Printing, page 94.

Jones and Bartlett, NFPA, *Fundamentals of Fire Fighter Skills*, 2nd Edition, 1st Printing, page 791.

Answer: A

15. Reference: NFPA 1001, 5.1.1

Delmar, *Firefighter's Handbook*, 3rd Edition, 1st Printing, page 500.

IFSTA, *Essentials of Fire Fighting and Fire Department Operations*, 5th Edition, 1st Printing, page 272.

Jones and Bartlett, NFPA, *Fundamentals of Fire Fighter Skills*, 2nd Edition, 1st Printing, page 248.

Answer: A

16. Reference: NFPA 1001, 5.1.1

Delmar, *Firefighter's Handbook*, 3rd Edition, 1st Printing, page 160.

IFSTA, *Essentials of Fire Fighting and Fire Department Operations*, 5th Edition, 1st Printing, page 180.

Jones and Bartlett, NFPA, *Fundamentals of Fire Fighter Skills*, 2nd Edition, 1st Printing, page 47.

Answer: B

17. Reference: NFPA 1001, 5.2.1 and 5.2.1(A)(B)

Delmar, *Firefighter's Handbook*, 3rd Edition, 1st Printing, page 76.

IFSTA, *Essentials of Fire Fighting and Fire Department Operations*, 5th Edition, 1st Printing, page 945.

Jones and Bartlett, NFPA, *Fundamentals of Fire Fighter Skills*, 2nd Edition, 1st Printing, page 96.

Answer: B

18. Reference: NFPA 1001, 5.2.1, 5.2.1(A)(B), 5.2.2, 5.2.2(A)(B), 5.2.3, and 5.2.3(A)(B)

Delmar, *Firefighter's Handbook*, 3rd Edition, 1st Printing, page 55.

IFSTA, *Essentials of Fire Fighting and Fire Department Operations*, 5th Edition, 1st Printing, page 927.

Jones and Bartlett, NFPA, *Fundamentals of Fire Fighter Skills*, 2nd Edition, 1st Printing, page 82.

Answer: A

19. Reference: NFPA 1001, 5.2.3 and 5.2.3(A)(B)

Delmar, *Firefighter's Handbook*, 3rd Edition, 1st Printing, page 73.

IFSTA, *Essentials of Fire Fighting and Fire Department Operations*, 5th Edition, 1st Printing, page 945.

Jones and Bartlett, NFPA, *Fundamentals of Fire Fighter Skills*, 2nd Edition, 1st Printing, page 93.

Answer: D

20. Reference: NFPA 1001, 5.3.1 and 5.3.1(A)(B)

Delmar, *Firefighter's Handbook*, 3rd Edition, 1st Printing, page 188.

IFSTA, *Essentials of Fire Fighting and Fire Department Operations*, 5th Edition, 1st Printing, page 203.

Jones and Bartlett, NFPA, *Fundamentals of Fire Fighter Skills*, 2nd Edition, 1st Printing, page 53.

Answer: B

21. Reference: NFPA 1001, 5.3.1, 5.3.1(A)(B), 5.3.5, 5.3.5(A)(B), 5.3.9, and 5.3.9(A)(B)
Delmar, *Firefighter's Handbook*, 3rd Edition, 1st Printing, page 177.
IFSTA, *Essentials of Fire Fighting and Fire Department Operations*, 5th Edition,
1st Printing, page 208.
Jones and Bartlett, NFPA, *Fundamentals of Fire Fighter Skills*, 2nd Edition,
1st Printing, page 52.
Answer: B

22. Reference: NFPA 1001, 5.3.1, 5.3.1(A)(B), 5.3.5, 5.3.5(A)(B), 5.3.10, 5.3.10(A),
5.3.11, and 5.3.11(A)
Delmar, *Firefighter's Handbook*, 3rd Edition, 1st Printing, page 162.
IFSTA, *Essentials of Fire Fighting and Fire Department Operations*, 5th Edition,
1st Printing, page 181.
Jones and Bartlett, NFPA, *Fundamentals of Fire Fighter Skills*, 2nd Edition,
1st Printing, page 48.
Answer: C

23. Reference: NFPA 1001, 5.3.1, 5.3.1(A)(B), 5.3.5, and 5.3.5(A)(B)
Delmar, *Firefighter's Handbook*, 3rd Edition, 1st Printing, page 162.
IFSTA, *Essentials of Fire Fighting and Fire Department Operations*, 5th Edition,
1st Printing, page 180.
Jones and Bartlett, NFPA, *Fundamentals of Fire Fighter Skills*, 2nd Edition,
1st Printing, pages 47–48.
Answer: A

24. Reference: NFPA 1001, 5.3.1, 5.3.1(A)(B), 5.3.5, 5.3.5(A)(B), 5.3.9 and 5.3.9(A)(B)
Delmar, *Firefighter's Handbook*, 3rd Edition, 1st Printing, pages 170–171.
IFSTA, *Essentials of Fire Fighting and Fire Department Operations*, 5th Edition,
1st Printing, page 210.
Jones and Bartlett, NFPA, *Fundamentals of Fire Fighter Skills*, 2nd Edition,
1st Printing, page 53.
Answer: D

25. Reference: NFPA 1001, 5.3.1, 5.3.1(A)(B), 5.3.5, 5.3.5(A)(B), 5.3.9, and 5.3.9(A)(B)
Delmar, *Firefighter's Handbook*, 3rd Edition, 1st Printing, page 177.
IFSTA, *Essentials of Fire Fighting and Fire Department Operations*, 5th Edition,
1st Printing, page 192.
Jones and Bartlett, NFPA, *Fundamentals of Fire Fighter Skills*, 2nd Edition,
1st Printing, page 52.
Answer: B

26. Reference: NFPA 1001, 5.3.1, 5.3.1(A), 5.3.5, 5.3.5(A)(B), 5.3.9, and 5.3.9(A)(B)
Delmar, *Firefighter's Handbook*, 3rd Edition, 1st Printing, pages 176–177.
IFSTA, *Essentials of Fire Fighting and Fire Department Operations*, 5th Edition,
1st Printing, page 192.
Jones and Bartlett, NFPA, *Fundamentals of Fire Fighter Skills*, 2nd Edition,
1st Printing, page 52.
Answer: C

27. Reference: NFPA 1001, 5.3.1 and 5.3.1(A)

Delmar, *Firefighter's Handbook*, 3rd Edition, 1st Printing, pages 162 and 1110.

IFSTA, *Essentials of Fire Fighting and Fire Department Operations*, 5th Edition, 1st Printing, page 180.

Jones and Bartlett, NFPA, *Fundamentals of Fire Fighter Skills*, 2nd Edition, 1st Printing, page 131.

Answer: C

28. Reference: NFPA 1001, 5.3.1, 5.3.1(A), 5.3.5, 5.3.5(A)(B), 5.3.11, and 5.3.11(A)

Delmar, *Firefighter's Handbook*, 3rd Edition, 1st Printing, page 161.

IFSTA, *Essentials of Fire Fighting and Fire Department Operations*, 5th Edition, 1st Printing, page 182.

Jones and Bartlett, NFPA, *Fundamentals of Fire Fighter Skills*, 2nd Edition, 1st Printing, page 131.

Answer: A

29. Reference: NFPA 1001, 5.3.1 and 5.3.1(A)(B)

Delmar, *Firefighter's Handbook*, 3rd Edition, 1st Printing, page 172.

IFSTA, *Essentials of Fire Fighting and Fire Department Operations*, 5th Edition, 1st Printing, pages 167 and 190.

Jones and Bartlett, NFPA, *Fundamentals of Fire Fighter Skills*, 2nd Edition, 1st Printing, page 49.

Answer: C

30. Reference: NFPA 1001, 5.3.1 and 5.3.1(A)(B)

Delmar, *Firefighter's Handbook*, 3rd Edition, 1st Printing, pages 180, 182, and 184.

IFSTA, *Essentials of Fire Fighting and Fire Department Operations*, 5th Edition, 1st Printing, page 200.

Jones and Bartlett, NFPA, *Fundamentals of Fire Fighter Skills*, 2nd Edition, 1st Printing, page 54.

Answer: A

31. Reference: NFPA 1001, 5.3.1, 5.3.1(A)(B), 5.1.1

Delmar, *Firefighter's Handbook*, 3rd Edition, 1st Printing, page 161.

IFSTA, *Essentials of Fire Fighting and Fire Department Operations*, 5th Edition, page 199.

Jones and Bartlett, NFPA, *Fundamentals of Fire Fighter Skills*, 2nd Edition, 1st Printing, page 40.

Answer: C

32. Reference: NFPA 1001, 5.3.1 and 5.3.1(A)

Delmar, *Firefighter's Handbook*, 3rd Edition, 1st Printing, page 162.

IFSTA, *Essentials of Fire Fighting and Fire Department Operations*, 5th Edition, 1st Printing, page 104.

Jones and Bartlett, NFPA, *Fundamentals of Fire Fighter Skills*, 2nd Edition, 1st Printing, page 48.

Answer: D

33. Reference: NFPA 1001, 5.3.1 and 5.3.1(A)(B)

Delmar, *Firefighter's Handbook*, 3rd Edition, 1st Printing, page 180.

IFSTA, *Essentials of Fire Fighting and Fire Department Operations*, 5th Edition, 1st Printing, page 190.

Jones and Bartlett, NFPA, *Fundamentals of Fire Fighter Skills*, 2nd Edition, 1st Printing, page 54.

Answer: D

34. Reference: NFPA 1001, 5.3.1 and 5.3.1(A)(B)

Delmar, *Firefighter's Handbook*, 3rd Edition, 1st Printing, page 179.

IFSTA, *Essentials of Fire Fighting and Fire Department Operations*, 5th Edition, 1st Printing, page 200.

Jones and Bartlett, NFPA, *Fundamentals of Fire Fighter Skills*, 2nd Edition, 1st Printing, page 56.

Answer: A

35. Reference: NFPA 1001, 5.3.2 and 5.3.2(A)(B)

Delmar, *Firefighter's Handbook*, 3rd Edition, 1st Printing, page 149.

IFSTA, *Essentials of Fire Fighting and Fire Department Operations*, 5th Edition, 1st Printing, pages 168 and 179.

Jones and Bartlett, NFPA, *Fundamentals of Fire Fighter Skills*, 2nd Edition, 1st Printing, page 43.

Answer: A

36. Reference: NFPA 1001, 5.3.2, 5.3.2(A), 5.3.3 and 5.3.3(A)(B)

Delmar, *Firefighter's Handbook*, 3rd Edition, 1st Printing, page 142.

IFSTA, *Essentials of Fire Fighting and Fire Department Operations*, 5th Edition, 1st Printing, page 175.

Jones and Bartlett, NFPA, *Fundamentals of Fire Fighter Skills*, 2nd Edition, 1st Printing, page 38.

Answer: A

37. Reference: NFPA 1001, 5.3.4 and 5.3.4(A)(B)

Delmar, *Firefighter's Handbook*, 3rd Edition, 1st Printing, page 600.

IFSTA, *Essentials of Fire Fighting and Fire Department Operations*, 5th Edition, 1st Printing, page 433.

Jones and Bartlett, NFPA, *Fundamentals of Fire Fighter Skills*, 2nd Edition, 1st Printing, page 305.

Answer: B

38. Reference: NFPA 1001, 5.3.4 and 5.3.4(A)

Delmar, *Firefighter's Handbook*, 3rd Edition, 1st Printing, page 601.

IFSTA, *Essentials of Fire Fighting and Fire Department Operations*, 5th Edition, 1st Printing, page 417.

Jones and Bartlett, NFPA, *Fundamentals of Fire Fighter Skills*, 2nd Edition, 1st Printing, page 301.

Answer: C

39. Reference: NFPA 1001, 5.3.4 and 5.3.4(A)(B)

Delmar, *Firefighter's Handbook*, 3rd Edition, 1st Printing, page 602.

IFSTA, *Essentials of Fire Fighting and Fire Department Operations*, 5th Edition, 1st Printing, page 423.

Jones and Bartlett, NFPA, *Fundamentals of Fire Fighter Skills*, 2nd Edition, 1st Printing, page 302.

Answer: D

40. Reference: NFPA 1001, 5.3.4 and 5.3.4(A)(B)

Delmar, *Firefighter's Handbook*, 3rd Edition, 1st Printing, pages 616–617.

IFSTA, *Essentials of Fire Fighting and Fire Department Operations*, 5th Edition, 1st Printing, page 434.

Jones and Bartlett, NFPA, *Fundamentals of Fire Fighter Skills*, 2nd Edition, 1st Printing, page 296.

Answer: A

41. Reference: NFPA 1001, 5.3.4 and 5.3.4(A)(B)

Delmar, *Firefighter's Handbook*, 3rd Edition, 1st Printing, pages 589–590.

IFSTA, *Essentials of Fire Fighting and Fire Department Operations*, 5th Edition, 1st Printing, page 446.

Jones and Bartlett, NFPA, *Fundamentals of Fire Fighter Skills*, 2nd Edition, 1st Printing, page 294.

Answer: C

42. Reference: NFPA 1001, 5.3.4 and 5.3.4(A)(B)

Delmar, *Firefighter's Handbook*, 3rd Edition, 1st Printing, page 594.

IFSTA, *Essentials of Fire Fighting and Fire Department Operations*, 5th Edition, 1st Printing, page 405.

Jones and Bartlett, NFPA, *Fundamentals of Fire Fighter Skills*, 2nd Edition, 1st Printing, pages 215 and 218.

Answer: C

43. Reference: NFPA 1001, 5.3.4 and 5.3.4(A)

Delmar, *Firefighter's Handbook*, 3rd Edition, 1st Printing, page 590.

IFSTA, *Essentials of Fire Fighting and Fire Department Operations*, 5th Edition, 1st Printing, page 410.

Jones and Bartlett, NFPA, *Fundamentals of Fire Fighter Skills*, 2nd Edition, 1st Printing, pages 223 and 294.

Answer: D

44. Reference: NFPA 1001, 5.3.4 and 5.3.4(A)(B)

Delmar, *Firefighter's Handbook*, 3rd Edition, 1st Printing, page 591.

IFSTA, *Essentials of Fire Fighting and Fire Department Operations*, 5th Edition, 1st Printing, page 430.

Jones and Bartlett, NFPA, *Fundamentals of Fire Fighter Skills*, 2nd Edition, 1st Printing, page 219.

Answer: C

45. Reference: NFPA 1001, 5.3.4 and 5.3.4(A)(B)
Delmar, *Firefighter's Handbook*, 3rd Edition, 1st Printing, page 597.
IFSTA, *Essentials of Fire Fighting and Fire Department Operations*, 5th Edition,
1st Printing, page 562.
Jones and Bartlett, NFPA, *Fundamentals of Fire Fighter Skills*, 2nd Edition,
1st Printing, page 427.
Answer: B

46. Reference: NFPA 1001, 5.3.4 and 5.3.4(A)
Delmar, *Firefighter's Handbook*, 3rd Edition, 1st Printing, pages 594–595.
IFSTA, *Essentials of Fire Fighting and Fire Department Operations*, 4th Edition,
1st Printing, page 256.
Jones and Bartlett, NFPA, *Fundamentals of Fire Fighter Skills*, 2nd Edition,
1st Printing, page 227.
Answer: C

47. Reference: NFPA 1001, 5.3.4 and 5.3.4(A)(B)
Delmar, *Firefighter's Handbook*, 3rd Edition, 1st Printing, page 587.
IFSTA, *Essentials of Fire Fighting and Fire Department Operations*, 5th Edition,
1st Printing, page 407.
Jones and Bartlett, NFPA, *Fundamentals of Fire Fighter Skills*, 2nd Edition,
1st Printing, page 293.
Answer: A

48. Reference: NFPA 1001, 5.3.5, 5.3.5(A)(B), 5.3.9, and 5.3.9(A)(B)
Delmar, *Firefighter's Handbook*, 3rd Edition, 1st Printing, page 520.
IFSTA, *Essentials of Fire Fighting and Fire Department Operations*, 5th Edition,
1st Printing, pages 308–309.
Jones and Bartlett, NFPA, *Fundamentals of Fire Fighter Skills*, 2nd Edition,
1st Printing, page 367.
Answer: D

49. Reference: NFPA 1001, 5.3.5 and 5.3.5(A)
Delmar, *Firefighter's Handbook*, 3rd Edition, 1st Printing, page 836.
IFSTA, *Essentials of Fire Fighting and Fire Department Operations*, 5th Edition,
1st Printing, page 71.
Jones and Bartlett, NFPA, *Fundamentals of Fire Fighter Skills*, 2nd Edition,
1st Printing, page 535.
Answer: D

50. Reference: NFPA 1001, 5.3.5 and 5.3.5(A)(B)
Delmar, *Firefighter's Handbook*, 3rd Edition, 1st Printing, page 121.
IFSTA, *Essentials of Fire Fighting and Fire Department Operations*, 5th Edition,
1st Printing, pages 77–78.
Jones and Bartlett, NFPA, *Fundamentals of Fire Fighter Skills*, 2nd Edition,
1st Printing, page 378.
Answer: A

51. References: NFPA 1001, 5.3.6 and 5.3.6(A)

Delmar, *Firefighter's Handbook*, 3rd Edition, 1st Printing, page 417.

IFSTA, *Essentials of Fire Fighting and Fire Department Operations*, 5th Edition, 1st Printing, page 475.

Jones and Bartlett, NFPA, *Fundamentals of Fire Fighter Skills*, 2nd Edition, 1st Printing, page 328.

Answer: A

52. References: NFPA 1001, 5.3.6, 5.3.6(A)(B), 5.3.11, 5.3.11(A)(B) 5.3.12, and 5.3.12(A)(B)

Delmar, *Firefighter's Handbook*, 3rd Edition, 1st Printing, pages 427 and 442.

IFSTA, *Essentials of Fire Fighting and Fire Department Operations*, 5th Edition, 1st Printing, pages 483 and 494.

Jones and Bartlett, NFPA, *Fundamentals of Fire Fighter Skills*, 2nd Edition, 1st Printing, pages 337 and 347.

Answer: B

53. References: NFPA 1001, 5.3.6 and 5.3.6(A)

Delmar, *Firefighter's Handbook*, 3rd Edition, 1st Printing, page 419.

IFSTA, *Essentials of Fire Fighting and Fire Department Operations*, 5th Edition, 1st Printing, page 476.

Jones and Bartlett, NFPA, *Fundamentals of Fire Fighter Skills*, 2nd Edition, 1st Printing, page 330.

Answer: B

54. References: NFPA 1001, 5.3.6 and 5.3.6(A)(B)

Delmar, *Firefighter's Handbook*, 3rd Edition, 1st Printing, pages 427–428.

IFSTA, *Essentials of Fire Fighting and Fire Department Operations*, 5th Edition, 1st Printing, page 485.

Jones and Bartlett, NFPA, *Fundamentals of Fire Fighter Skills*, 2nd Edition, 1st Printing, page 339.

Answer: A

55. References: NFPA 1001, 5.3.6 and 5.3.6(A)(B)

Delmar, *Firefighter's Handbook*, 3rd Edition,1st Printing, page 432.

IFSTA, *Essentials of Fire Fighting and Fire Department Operations*, 5th Edition, 1st Printing, pages 504–505.

Jones and Bartlett, NFPA, *Fundamentals of Fire Fighter Skills*, 2nd Edition, 1st Printing, page 356.

Answer: C

56. References: NFPA 1001, 5.3.6 and 5.3.6(A)(B)

Delmar, *Firefighter's Handbook*, 3rd Edition, 1st Printing, pages 442, 455, and 456.

IFSTA, *Essentials of Fire Fighting and Fire Department Operations*, 5th Edition, 1st Printing, pages 515–517.

Jones and Bartlett, NFPA, *Fundamentals of Fire Fighter Skills*, 2nd Edition, 1st Printing, page 347.

Answer: C

57. References: NFPA 1001, 5.3.6 and 5.3.6(A)

Delmar, *Firefighter's Handbook*, 3rd Edition, 1st Printing, page 419.

IFSTA, *Essentials of Fire Fighting and Fire Department Operations*, 5th Edition, 1st Printing, page 476.

Jones and Bartlett, NFPA, *Fundamentals of Fire Fighter Skills*, 2nd Edition, 1st Printing, page 330.

Answer: B

58. References: NFPA 1001, 5.3.6 and 5.3.6(A)(B)

Delmar, *Firefighter's Handbook*, 3rd Edition, 1st Printing, page 443.

IFSTA, *Essentials of Fire Fighting and Fire Department Operations*, 5th Edition, 1st Printing, pages 498–499, and 521–524.

Jones and Bartlett, NFPA, *Fundamentals of Fire Fighter Skills*, 2nd Edition, 1st Printing, page 347.

Answer: D

59. Reference: NFPA 1001, 5.3.6 and 5.3.6(A)

Delmar, *Firefighter's Handbook*, 3rd Edition, 1st Printing, page 417.

IFSTA, *Essentials of Fire Fighting and Fire Department Operations*, 5th Edition, 1st Printing, pages 472–473.

Jones and Bartlett, NFPA, *Fundamentals of Fire Fighter Skills*, 2nd Edition, 1st Printing, page 328.

Answer: D

60. Reference: NFPA 1001, 5.3.6 and 5.3.6(A)(B)

Delmar, *Firefighter's Handbook*, 3rd Edition, 1st Printing, page 417.

IFSTA, *Essentials of Fire Fighting and Fire Department Operations*, 5th Edition, 1st Printing, page 475.

Jones and Bartlett, NFPA, *Fundamentals of Fire Fighter Skills*, 2nd Edition, 1st Printing, page 329.

Answer: A

61. Reference: NFPA 1001, 5.3.6 and 5.3.6(A)(B)

Delmar, *Firefighter's Handbook*, 3rd Edition, 1st Printing, pages 439 and 442.

IFSTA, *Essentials of Fire Fighting and Fire Department Operations*, 5th Edition, 1st Printing, page 489.

Jones and Bartlett, NFPA, *Fundamentals of Fire Fighter Skills*, 2nd Edition, 1st Printing, page 341.

Answer: C

62. Reference: NFPA 1001, 5.3.8, 5.3.8(A), 5.3.10, and 5.3.10(A)

Delmar, *Firefighter's Handbook*, 3rd Edition, 1st Printing, page 261.

IFSTA, *Essentials of Fire Fighting and Fire Department Operations*, 5th Edition, 1st Printing, page 652.

Jones and Bartlett, NFPA, *Fundamentals of Fire Fighter Skills*, 2nd Edition, 1st Printing, pages 618–619.

Answer: C

63. Reference: NFPA 1001, 5.3.8, 5.3.8(A), 5.3.10, and 5.3.10(A)

Delmar, *Firefighter's Handbook*, 3rd Edition, 1st Printing, pages 255–256.

IFSTA, *Essentials of Fire Fighting and Fire Department Operations*, 5th Edition, 1st Printing, page 632.

Jones and Bartlett, NFPA, *Fundamentals of Fire Fighter Skills*, 2nd Edition, 1st Printing, page 466.

Answer: C

64. Reference: NFPA 1001, 5.3.9 and 5.3.9(A)

Delmar, *Firefighter's Handbook*, 3rd Edition, 1st Printing, page 519.

IFSTA, *Essentials of Fire Fighting and Fire Department Operations*, 5th Edition, 1st Printing, page 305.

Jones and Bartlett, NFPA, *Fundamentals of Fire Fighter Skills*, 2nd Edition, 1st Printing, page 366.

Answer: A

65. Reference: NFPA 1001, 5.3.9 and 5.3.9(A)(B)

Delmar, *Firefighter's Handbook*, 3rd Edition, 1st Printing, page 526.

IFSTA, *Essentials of Fire Fighting and Fire Department Operations*, 5th Edition, 1st Printing, page 310.

Jones and Bartlett, NFPA, *Fundamentals of Fire Fighter Skills*, 2nd Edition, 1st Printing, pages 370 and 374.

Answer: C

66. Reference: NFPA 1001, 5.3.9 and 5.3.9(A)(B)

Delmar, *Firefighter's Handbook*, 3rd Edition, 1st Printing, page 524.

IFSTA, *Essentials of Fire Fighting and Fire Department Operations*, 5th Edition, 1st Printing, pages 309–310.

Jones and Bartlett, NFPA, *Fundamentals of Fire Fighter Skills*, 2nd Edition, 1st Printing, page 370.

Answer: A

67. Reference: NFPA 1001, 5.3.9 and 5.3.9(A)(B)

Delmar, *Firefighter's Handbook*, 3rd Edition, 1st Printing, pages 521–522.

IFSTA, *Essentials of Fire Fighting and Fire Department Operations*, 5th Edition, 1st Printing, page 316.

Jones and Bartlett, NFPA, *Fundamentals of Fire Fighter Skills*, 2nd Edition, 1st Printing, page 371.

Answer: B

68. Reference: NFPA 1001, 5.3.9 and 5.3.9(A)(B)

Delmar, *Firefighter's Handbook*, 3rd Edition, 1st Printing, page 837.

IFSTA, *Essentials of Fire Fighting and Fire Department Operations*, 5th Edition, 1st Printing, pages 320–321.

Jones and Bartlett, NFPA, *Fundamentals of Fire Fighter Skills*, 2nd Edition, 1st Printing, pages 538–539.

Answer: D

69. Reference: NFPA 1001, 5.3.9 and 5.3.9(A)(B)

Delmar, *Firefighter's Handbook*, 3rd Edition, 1st Printing, pages 524 and 526.

IFSTA, *Essentials of Fire Fighting and Fire Department Operations*, 5th Edition, 1st Printing, pages 309–310.

Jones and Bartlett, NFPA, *Fundamentals of Fire Fighter Skills*, 2nd Edition, 1st Printing, pages 370–374.

Answer: A

70. Reference: NFPA 1001, 5.3.9, and 5.3.9(A)(B)

Delmar, *Firefighter's Handbook*, 3rd Edition, 1st Printing, page 833.

IFSTA, *Essentials of Fire Fighting and Fire Department Operations*, 5th Edition, 1st Printing, page 307.

Jones and Bartlett, NFPA, *Fundamentals of Fire Fighter Skills*, 2nd Edition, 1st Printing, page 536.

Answer: C

71. Reference: NFPA 1001, 5.3.9 and 5.3.9(A)(B)

Delmar, *Firefighter's Handbook*, 3rd Edition, 1st Printing, page 540.

IFSTA, *Essentials of Fire Fighting and Fire Department Operations*, 5th Edition, 1st Printing, page 330.

Jones and Bartlett, NFPA, *Fundamentals of Fire Fighter Skills*, 2nd Edition, 1st Printing, page 391.

Answer: C

72. Reference: NFPA 1001, 5.3.10, 5.3.10(A), 5.3.1 and 5.3.1(A)

Delmar, *Firefighter's Handbook*, 3rd Edition, 1st Printing, page 160.

IFSTA, *Essentials of Fire Fighting and Fire Department Operations*, 5th Edition, 1st Printing, page 180.

Jones and Bartlett, NFPA, *Fundamentals of Fire Fighter Skills*, 2nd Edition, 1st Printing, page 131.

Answer: A

73. Reference: NFPA 1001, 5.3.10, 5.3.10(A), 5.3.11, and 5.3.11(A)

Delmar, *Firefighter's Handbook*, 3rd Edition, 1st Printing, page 911.

IFSTA, *Essentials of Fire Fighting and Fire Department Operations*, 5th Edition, 1st Printing, page 105.

Jones and Bartlett, NFPA, *Fundamentals of Fire Fighter Skills*, 2nd Edition, 1st Printing, page 142.

Answer: C

74. Reference: NFPA 1001, 5.3.10 and 5.3.10(A)(B)

Delmar, *Firefighter's Handbook*, 3rd Edition, 1st Printing, pages 377 and 705.

IFSTA, *Essentials of Fire Fighting and Fire Department Operations*, 5th Edition, 1st Printing, page 674.

Jones and Bartlett, NFPA, *Fundamentals of Fire Fighter Skills*, 2nd Edition, 1st Printing, page 514.

Answer: A

75. Reference: NFPA 1001, 5.3.10, 5.3.10(A)(B), 5.3.8 and 5.3.8(A)(B)

Delmar, *Firefighter's Handbook*, 3rd Edition, 1st Printing, pages 326–328.

IFSTA, *Essentials of Fire Fighting and Fire Department Operations*, 5th Edition, 1st Printing, page 728.

Jones and Bartlett, NFPA, *Fundamentals of Fire Fighter Skills*, 2nd Edition, 1st Printing, page 619.

Answer: B

76. Reference: NFPA 1001, 5.3.10, 5.3.10(A)(B), 5.3.8 and 5.3.8(A)(B)

Delmar, *Firefighter's Handbook*, 3rd Edition, 1st Printing, pages 326–327.

IFSTA, *Essentials of Fire Fighting and Fire Department Operations*, 5th Edition, 1st Printing, page 726.

Jones and Bartlett, NFPA, *Fundamentals of Fire Fighter Skills*, 2nd Edition, 1st Printing, page 619.

Answer: A

77. Reference: NFPA 1001, 5.3.10, 5.3.10(A)(B), 5.3.8 and 5.3.8(A)(B)

Delmar, *Firefighter's Handbook*, 3rd Edition, 1st Printing, pages 299–300.

IFSTA, *Essentials of Fire Fighting and Fire Department Operations*, 5th Edition, 1st Printing, pages 675–676.

Jones and Bartlett, NFPA, *Fundamentals of Fire Fighter Skills*, 2nd Edition, 1st Printing, page 511.

Answer: C

78. Reference: NFPA 1001, 5.3.10, 5.3.10(A)(B), 5.3.8 and 5.3.8(A)(B)

Delmar, *Firefighter's Handbook*, 3rd Edition, 1st Printing, page 261.

IFSTA, *Essentials of Fire Fighting and Fire Department Operations*, 5th Edition, 1st Printing, page 649.

Jones and Bartlett, NFPA, *Fundamentals of Fire Fighter Skills*, 2nd Edition, 1st Printing, page 481.

Answer: C

79. Reference: NFPA 1001, 5.3.10, 5.3.10(A)(B), 5.3.8 and 5.3.8(A)(B)

Delmar, *Firefighter's Handbook*, 3rd Edition, 1st Printing, pages 326–327.

IFSTA, *Essentials of Fire Fighting and Fire Department Operations*, 5th Edition, 1st Printing, page 728.

Jones and Bartlett, NFPA, *Fundamentals of Fire Fighter Skills*, 2nd Edition, 1st Printing, pages 517 and 619.

Answer: B

80. Reference: NFPA 1001, 5.3.10, 5.3.10(A)(B), 5.3.8 and 5.3.8(A)(B)

Delmar, *Firefighter's Handbook*, 3rd Edition, 1st Printing, pages 264 and 266.

IFSTA, *Essentials of Fire Fighting and Fire Department Operations*, 5th Edition, 1st Printing, page 688.

Jones and Bartlett, NFPA, *Fundamentals of Fire Fighter Skills*, 2nd Edition, 1st Printing, pages 470 and 472.

Answer: A

81. Reference: NFPA 1001, 5.3.10 and 5.3.10(A)(B)

Delmar, *Firefighter's Handbook*, 3rd Edition, 1st Printing, page 315.

IFSTA, *Essentials of Fire Fighting and Fire Department Operations*, 5th Edition, 1st Printing, pages 543–544.

Jones and Bartlett, NFPA, *Fundamentals of Fire Fighter Skills*, 2nd Edition, 1st Printing, page 618.

Answer: C

82. Reference: NFPA 1001, 5.3.10, 5.3.10(A)(B), 5.3.8 and 5.3.8(A)(B)

Delmar, *Firefighter's Handbook*, 3rd Edition, 1st Printing, page 273.

IFSTA, *Essentials of Fire Fighting and Fire Department Operations*, 5th Edition, 1st Printing, page 656.

Jones and Bartlett, NFPA, *Fundamentals of Fire Fighter Skills*, 2nd Edition, 1st Printing, page 493.

Answer: A

83. Reference: NFPA 1001, 5.3.10, 5.3.10(A)(B), 5.3.15, and 5.3.15(A)(B)

Delmar, *Firefighter's Handbook*, 3rd Edition, 1st Printing, page 304.

IFSTA, *Essentials of Fire Fighting and Fire Department Operations*, 5th Edition, 1st Printing, pages 665–666.

Jones and Bartlett, NFPA, *Fundamentals of Fire Fighter Skills*, 2nd Edition, 1st Printing, page 487.

Answer: B

84. Reference: NFPA 1001, 5.3.10, 5.3.10(A), 5.3.7, 5.3.7(A), 5.3.8 and 5.3.8(A)

Delmar, *Firefighter's Handbook*, 3rd Edition, 1st Printing, page 328.

IFSTA, *Essentials of Fire Fighting and Fire Department Operations*, 5th Edition, 1st Printing, page 731.

Jones and Bartlett, NFPA, *Fundamentals of Fire Fighter Skills*, 2nd Edition, 1st Printing, page 517.

Answer: C

85. Reference: NFPA 1001, 5.3.10 and 5.3.10(A)(B)

Delmar, *Firefighter's Handbook*, 3rd Edition, 1st Printing, page 258.

IFSTA, *Essentials of Fire Fighting and Fire Department Operations*, 5th Edition, 1st Printing, page 634.

Jones and Bartlett, NFPA, *Fundamentals of Fire Fighter Skills*, 2nd Edition, 1st Printing, page 466.

Answer: C

86. Reference: NFPA 1001, 5.3.11, 5.3.11(A), 5.3.12, 5.3.12(A), 5.3.10 and 5.3.10(A)(B)

Delmar, *Firefighter's Handbook*, 3rd Edition, 1st Printing, page 631.

IFSTA, *Essentials of Fire Fighting and Fire Department Operations*, 5th Edition, 1st Printing, page 541.

Jones and Bartlett, NFPA, *Fundamentals of Fire Fighter Skills*, 2nd Edition, 1st Printing, page 403.

Answer: D

87. Reference: NFPA 1001, 5.3.11, 5.3.11(A), 5.3.12 and 5.3.12(A)
Delmar, *Firefighter's Handbook*, 3rd Edition, 1st Printing, page 97.
IFSTA, *Essentials of Fire Fighting and Fire Department Operations*, 5th Edition, 1st Printing, page 95.
Jones and Bartlett, NFPA, *Fundamentals of Fire Fighter Skills*, 2nd Edition, 1st Printing, page 132.
Answer: D

88. Reference: NFPA 1001, 5.3.11, 5.3.11(A), 5.3.12, 5.3.12(A), 5.3.13 and 5.3.13(A)
Delmar, *Firefighter's Handbook*, 3rd Edition, 1st Printing, page 99.
IFSTA, *Essentials of Fire Fighting and Fire Department Operations*, 5th Edition, 1st Printing, page 95.
Jones and Bartlett, NFPA, *Fundamentals of Fire Fighter Skills*, 2nd Edition, 1st Printing, pages 132–133.
Answer: B

89. Reference: NFPA 1001, 5.3.11, 5.3.11(A), 5.3.12, 5.3.12(A), 5.3.13 and 5.3.13(A)
Delmar, *Firefighter's Handbook*, 3rd Edition, 1st Printing, page 101.
IFSTA, *Essentials of Fire Fighting and Fire Department Operations*, 5th Edition, 1st Printing, page 116.
Jones and Bartlett, NFPA, *Fundamentals of Fire Fighter Skills*, 2nd Edition, 1st Printing, pages 136–137.
Answer: C

90. Reference: NFPA 1001, 5.3.11, 5.3.11(A), 5.3.12, 5.3.12(A), 5.3.13 and 5.3.13(A)
Delmar, *Firefighter's Handbook*, 3rd Edition, 1st Printing, page 100.
IFSTA, *Essentials of Fire Fighting and Fire Department Operations*, 5th Edition, 1st Printing, page 114.
Jones and Bartlett, NFPA, *Fundamentals of Fire Fighter Skills*, 2nd Edition, 1st Printing, page 135.
Answer: A

91. Reference: NFPA 1001, 5.3.11, 5.3.11(A), 5.3.12, 5.3.12(A), 5.3.13 and 5.3.13(A)
Delmar, *Firefighter's Handbook*, 3rd Edition, 1st Printing, page 95.
IFSTA, *Essentials of Fire Fighting and Fire Department Operations*, 5th Edition, 1st Printing, page 97.
Jones and Bartlett, NFPA, *Fundamentals of Fire Fighter Skills*, 2nd Edition, 1st Printing, page 141.
Answer: B

92. Reference: NFPA 1001, 5.3.11, 5.3.11(A), 5.3.12, and 5.3.12(A)
Delmar, *Firefighter's Handbook*, 3rd Edition, 1st Printing, page 95.
IFSTA, *Essentials of Fire Fighting and Fire Department Operations*, 5th Edition, 1st Printing, page 97.
Jones and Bartlett, NFPA, *Fundamentals of Fire Fighter Skills*, 2nd Edition, 1st Printing, page 141.
Answer: D

93. Reference: NFPA 1001, 5.3.11, 5.3.11(A), 5.3.12, and 5.3.12(A)
Delmar, *Firefighter's Handbook*, 3rd Edition, 1st Printing, pages 105 and 112.
IFSTA, *Essentials of Fire Fighting and Fire Department Operations*, 5th Edition, 1st Printing, page 774.
Jones and Bartlett, NFPA, *Fundamentals of Fire Fighter Skills*, 2nd Edition, 1st Printing, page 142.
Answer: C

94. Reference: NFPA 1001, 5.3.11 and 5.3.11(A)(B)
Delmar, *Firefighter's Handbook*, 3rd Edition, 1st Printing, pages 645–646.
IFSTA, *Essentials of Fire Fighting and Fire Department Operations*, 5th Edition, 1st Printing, page 575.
Jones and Bartlett, NFPA, *Fundamentals of Fire Fighter Skills*, 2nd Edition, 1st Printing, page 416.
Answer: B

95. Reference: NFPA 1001, 5.3.12 and 5.3.12(A)(B)
Delmar, *Firefighter's Handbook*, 3rd Edition, 1st Printing, page 631.
IFSTA, *Essentials of Fire Fighting and Fire Department Operations*, 5th Edition, 1st Printing, page 541.
Jones and Bartlett, NFPA, *Fundamentals of Fire Fighter Skills*, 2nd Edition, 1st Printing, page 405.
Answer: D

96. Reference: NFPA 1001, 5.3.12, 5.3.12(A)(B), 5.3.11 and 5.3.11(A)(B)
Delmar, *Firefighter's Handbook*, 3rd Edition, 1st Printing, page 632.
IFSTA, *Essentials of Fire Fighting and Fire Department Operations*, 5th Edition, 1st Printing, page 550.
Jones and Bartlett, NFPA, *Fundamentals of Fire Fighter Skills*, 2nd Edition, 1st Printing, page 405.
Answer: C

97. Reference: NFPA 1001, 5.3.12 and 5.3.12(A)
Delmar, *Firefighter's Handbook*, 3rd Edition, 1st Printing, page 397.
IFSTA, *Essentials of Fire Fighting and Fire Department Operations*, 5th Edition, 1st Printing, page 148.
Jones and Bartlett, NFPA, *Fundamentals of Fire Fighter Skills*, 2nd Edition, 1st Printing, page 160.
Answer: C

98. Reference: NFPA 1001, 5.3.12, 5.3.12(A), 5.3.10, 5.3.10(A), 5.3.11 and 5.3.11(A)
Delmar, *Firefighter's Handbook*, 3rd Edition, 1st Printing, page 404.
IFSTA, *Essentials of Fire Fighting and Fire Department Operations*, 5th Edition, 1st Printing, page 156.
Jones and Bartlett, NFPA, *Fundamentals of Fire Fighter Skills*, 2nd Edition, 1st Printing, page 422.
Answer: A

99. Reference: NFPA 1001, 5.3.12 and 5.3.12(A)(B)
Delmar, *Firefighter's Handbook*, 3rd Edition, 1st Printing, pages 655 and 656.
IFSTA, *Essentials of Fire Fighting and Fire Department Operations*, 5th Edition, 1st Printing, pages 560–561.
Jones and Bartlett, NFPA, *Fundamentals of Fire Fighter Skills*, 2nd Edition, 1st Printing, page 425.
Answer: D

100. Reference: NFPA 1001, 5.3.12 and 5.3.12(A)
Delmar, *Firefighter's Handbook*, 3rd Edition, 1st Printing, page 398.
IFSTA, *Essentials of Fire Fighting and Fire Department Operations*, 5th Edition, 1st Printing, page 140.
Jones and Bartlett, NFPA, *Fundamentals of Fire Fighter Skills*, 2nd Edition, 1st Printing, page 162.
Answer: B

101. Reference: NFPA 1001, 5.3.12 and 5.3.12(A)
Delmar, *Firefighter's Handbook*, 3rd Edition, 1st Printing, page 648.
IFSTA, *Essentials of Fire Fighting and Fire Department Operations*, 5th Edition, 1st Printing, page 556.
Jones and Bartlett, NFPA, *Fundamentals of Fire Fighter Skills*, 2nd Edition, 1st Printing, page 411.
Answer: B

102. Reference: NFPA 1001, 5.3.12, 5.3.12(A), 5.3.10, and 5.3.10(A)(B)
Delmar, *Firefighter's Handbook*, 3rd Edition, 1st Printing, page 391.
IFSTA, *Essentials of Fire Fighting and Fire Department Operations*, 5th Edition, 1st Printing, page 140.
Jones and Bartlett, NFPA, *Fundamentals of Fire Fighter Skills*, 2nd Edition, 1st Printing, page 154.
Answer: C

103. Reference: NFPA 1001, 5.3.13 and 5.3.13(A)(B)
Delmar, *Firefighter's Handbook*, 3rd Edition, 1st Printing, page 743.
IFSTA, *Essentials of Fire Fighting and Fire Department Operations*, 5th Edition, 1st Printing, pages 869 and 881.
Jones and Bartlett, NFPA, *Fundamentals of Fire Fighter Skills*, 2nd Edition, 1st Printing, page 570.
Answer: D

104. Reference: NFPA 1001, 5.3.13 and 5.3.13(A)(B)
Delmar, *Firefighter's Handbook*, 3rd Edition, 1st Printing, page 743.
IFSTA, *Essentials of Fire Fighting and Fire Department Operations*, 5th Edition, 1st Printing, pages 869 and 881.
Jones and Bartlett, NFPA, *Fundamentals of Fire Fighter Skills*, 2nd Edition, 1st Printing, page 570.
Answer: A

105. Reference: NFPA 1001, 5.3.13 and 5.3.13(A)

Delmar, *Firefighter's Handbook*, 3rd Edition, 1st Printing, page 743.

IFSTA, *Essentials of Fire Fighting and Fire Department Operations*, 5th Edition, 1st Printing, pages 869 and 881.

Jones and Bartlett, NFPA, *Fundamentals of Fire Fighter Skills*, 2nd Edition, 1st Printing, page 575.

Answer: B

106. Reference: NFPA 1001, 5.3.13 and 5.3.13(A)(B)

Delmar, *Firefighter's Handbook*, 3rd Edition, 1st Printing, page 743.

IFSTA, *Essentials of Fire Fighting and Fire Department Operations* Skills, 5th Edition, 1st Printing, page 869.

Jones and Bartlett, NFPA, *Fundamentals of Fire Fighter Skills*, 2nd Edition, 1st Printing, page 575.

Answer: B

107. Reference: NFPA 1001, 5.3.14 and 5.3.14(A)

Delmar, *Firefighter's Handbook*, 3rd Edition, 1st Printing, pages 740–741.

IFSTA, *Essentials of Fire Fighting and Fire Department Operations*, 5th Edition, 1st Printing, pages 879, and 898–899.

Jones and Bartlett, NFPA, *Fundamentals of Fire Fighter Skills*, 2nd Edition, 1st Printing, page 564.

Answer: B

108. Reference: NFPA 1001, 5.3.14 and 5.3.14(A)(B)

Delmar, *Firefighter's Handbook*, 3rd Edition, 1st Printing, page 740.

IFSTA, *Essentials of Fire Fighting and Fire Department Operations*, 5th Edition, 1st Printing, page 870.

Jones and Bartlett, NFPA, *Fundamentals of Fire Fighter Skills*, 2nd Edition, 1st Printing, page 564.

Answer: B

109. Reference: NFPA 1001, 5.3.14 and 5.3.14(A)(B)

Delmar, *Firefighter's Handbook*, 3rd Edition, 1st Printing, pages 733, 738, and 740.

IFSTA, *Essentials of Fire Fighting and Fire Department Operations*, 5th Edition, 1st Printing, pages 879–880.

Jones and Bartlett, NFPA, *Fundamentals of Fire Fighter Skills*, 2nd Edition, 1st Printing, page 564.

Answer: A

110. Reference: NFPA 1001, 5.3.14 and 5.3.14(A)

Delmar, *Firefighter's Handbook*, 3rd Edition, 1st Printing, page 374.

IFSTA, *Essentials of Fire Fighting and Fire Department Operations*, 5th Edition, 1st Printing, page 850.

Jones and Bartlett, NFPA, *Fundamentals of Fire Fighter Skills*, 2nd Edition, 1st Printing, pages 950 and 957.

Answer: C

111. Reference: NFPA 1001, 5.3.14 and 5.3.14(A)(B)

Delmar, *Firefighter's Handbook*, 3rd Edition, 1st Printing, pages 368–369.

IFSTA, *Essentials of Fire Fighting and Fire Department Operations*, 5th Edition, 1st Printing, pages 856–857, 860–861.

Jones and Bartlett, NFPA, *Fundamentals of Fire Fighter Skills*, 2nd Edition, 1st Printing, page 561.

Answer: B

112. Reference: NFPA 1001, 5.3.14 and 5.3.14(A)

Delmar, *Firefighter's Handbook*, 3rd Edition, 1st Printing, page 361.

IFSTA, *Essentials of Fire Fighting and Fire Department Operations*, 5th Edition, 1st Printing, page 845.

Jones and Bartlett, NFPA, *Fundamentals of Fire Fighter Skills*, 1st Edition, 1st Printing, page 954.

Answer: D

113. Reference: NFPA 1001, 5.3.14 and 5.3.14(A)(B)

Delmar, *Firefighter's Handbook*, 3rd Edition, 1st Printing, page 738.

IFSTA, *Essentials of Fire Fighting and Fire Department Operations*, 5th Edition, page 870.

Jones and Bartlett, NFPA, *Fundamentals of Fire Fighter Skills*, 2nd Edition, 1st Printing, page 565.

Answer: C

114. Reference: NFPA 1001, 5.3.14 and 5.3.14(A)(B)

Delmar, *Firefighter's Handbook*, 3rd Edition, 1st Printing, page 729.

IFSTA, *Essentials of Fire Fighting and Fire Department Operations*, 5th Edition, 1st Printing, page 849.

Jones and Bartlett, NFPA, *Fundamentals of Fire Fighter Skills*, 2nd Edition, 1st Printing, page 565.

Answer: B

115. Reference: NFPA 1001, 5.3.15 and 5.3.15(A)

Delmar, *Firefighter's Handbook*, 3rd Edition, 1st Printing, page 256.

IFSTA, *Essentials of Fire Fighting and Fire Department Operations*, 5th Edition, 1st Printing, page 633.

Jones and Bartlett, NFPA, *Fundamentals of Fire Fighter Skills*, 2nd Edition, 1st Printing, page 474.

Answer: A

116. Reference: NFPA 1001, 5.3.15 and 5.3.15(A)(B)

Delmar, *Firefighter's Handbook*, 3rd Edition, 1st Printing, page 236.

IFSTA, *Essentials of Fire Fighting and Fire Department Operations*, 5th Edition, 1st Printing, page 612.

Jones and Bartlett, NFPA, *Fundamentals of Fire Fighter Skills*, 2nd Edition, 1st Printing, pages 445–446.

Answer: C

117. Reference: NFPA 1001, 5.3.15 and 5.3.15(A)

Delmar, *Firefighter's Handbook*, 3rd Edition, 1st Printing, page 238.

IFSTA, *Essentials of Fire Fighting and Fire Department Operations*, 5th Edition, 1st Printing, pages 605–607.

Jones and Bartlett, NFPA, *Fundamentals of Fire Fighter Skills*, 2nd Edition, 1st Printing, page 452.

Answer: A

118. Reference: NFPA 1001, 5.3.15 and 5.3.15(A)(B)

Delmar, *Firefighter's Handbook*, 3rd Edition, 1st Printing, pages 239–240.

IFSTA, *Essentials of Fire Fighting and Fire Department Operations*, 5th Edition, 1st Printing, page 611.

Jones and Bartlett, NFPA, *Fundamentals of Fire Fighter Skills*, 2nd Edition, 1st Printing, page 453.

Answer: C

119. Reference: NFPA 1001, 5.3.15 and 5.3.15(A)

Delmar, *Firefighter's Handbook*, 3rd Edition, 1st Printing, page 247.

IFSTA, *Essentials of Fire Fighting and Fire Department Operations*, 5th Edition, 1st Printing, page 604.

Jones and Bartlett, NFPA, *Fundamentals of Fire Fighter Skills*, 2nd Edition, 1st Printing, page 457.

Answer: B

120. Reference: NFPA 1001, 5.3.15 and 5.3.15(A)

Delmar, *Firefighter's Handbook*, 3rd Edition, 1st Printing, page 239.

IFSTA, *Essentials of Fire Fighting and Fire Department Operations*, 5th Edition, 1st Printing, pages 605–606.

Jones and Bartlett, NFPA, *Fundamentals of Fire Fighter Skills*, 2nd Edition, 1st Printing, page 452.

Answer: C

121. Reference: NFPA 1001, 5.3.16 and 5.3.16(A)(B)

Delmar, *Firefighter's Handbook*, 3rd Edition, 1st Printing, page 215.

IFSTA, *Essentials of Fire Fighting and Fire Department Operations*, 5th Edition, 1st Printing, page 246.

Jones and Bartlett, NFPA, *Fundamentals of Fire Fighter Skills*, 2nd Edition, 1st Printing, pages 183–184.

Answer: C

122. Reference: NFPA 1001, 5.3.16 and 5.3.16(A)(B)

Delmar, *Firefighter's Handbook*, 3rd Edition, 1st Printing, pages 223–224.

IFSTA, *Essentials of Fire Fighting and Fire Department Operations*, 5th Edition, 1st Printing, page 247.

Jones and Bartlett, NFPA, *Fundamentals of Fire Fighter Skills*, 2nd Edition, 1st Printing, page 184.

Answer: B

123. Reference: NFPA 1001, 5.3.16 and 5.3.16(A)(B)

Delmar, *Firefighter's Handbook*, 3rd Edition, 1st Printing, page 218.

IFSTA, *Essentials of Fire Fighting and Fire Department Operations*, 5th Edition, 1st Printing, pages 249–250.

Jones and Bartlett, NFPA, *Fundamentals of Fire Fighter Skills*, 2nd Edition, 1st Printing, page 188.

Answer: B

124. Reference: NFPA 1001, 5.3.16 and 5.3.16(A)

Delmar, *Firefighter's Handbook*, 3rd Edition, 1st Printing, page 219.

IFSTA, *Essentials of Fire Fighting and Fire Department Operations*, 5th Edition, 1st Printing, page 249.

Jones and Bartlett, NFPA, *Fundamentals of Fire Fighter Skills*, 2nd Edition, 1st Printing, page 184.

Answer: A

125. Reference: NFPA 1001, 5.3.16 and 5.3.16(A)(B)

Delmar, *Firefighter's Handbook*, 3rd Edition, 1st Printing, page 219.

IFSTA, *Essentials of Fire Fighting and Fire Department Operations*, 5th Edition, 1st Printing, page 249.

Jones and Bartlett, NFPA, *Fundamentals of Fire Fighter Skills*, 2nd Edition, 1st Printing, page 184.

Answer: C

126. Reference: NFPA 1001, 5.3.16 and 5.3.16(A)(B)

Delmar, *Firefighter's Handbook*, 3rd Edition, 1st Printing, page 219.

IFSTA, *Essentials of Fire Fighting and Fire Department Operations*, 5th Edition, 1st Printing, page 249.

Jones and Bartlett, NFPA, *Fundamentals of Fire Fighter Skills*, 2nd Edition, 1st Printing, page 184.

Answer: D

127. Reference: NFPA 1001, 5.3.16 and 5.3.16(A)(B)

Delmar, *Firefighter's Handbook*, 3rd Edition, 1st Printing, page 189.

IFSTA, *Essentials of Fire Fighting and Fire Department Operations*, 5th Edition, 1st Printing, page 249.

Jones and Bartlett, NFPA, *Fundamentals of Fire Fighter Skills*, 2nd Edition, 1st Printing, page 184.

Answer: B

128. Reference: NFPA 1001, 5.3.16 and 5.3.16(A)(B)

Delmar, *Firefighter's Handbook*, 3rd Edition, 1st Printing, page 216.

IFSTA, *Essentials of Fire Fighting and Fire Department Operations*, 5th Edition, 1st Printing, page 111.

Jones and Bartlett, NFPA, *Fundamentals of Fire Fighter Skills*, 2nd Edition, 1st Printing, page 183.

Answer: D

129. Reference: NFPA 1001, 5.3.16 and 5.3.16(A)(B)

Delmar, *Firefighter's Handbook*, 3rd Edition, 1st Printing, page 218.

IFSTA, *Essentials of Fire Fighting and Fire Department Operations*, 5th Edition, 1st Printing, page 241.

Jones and Bartlett, NFPA, *Fundamentals of Fire Fighter Skills*, 2nd Edition, 1st Printing, page 189.

Answer: B

130. Reference: NFPA 1001, 5.3.16 and 5.3.16(A)(B)

Delmar, *Firefighter's Handbook*, 3rd Edition, 1st Printing, page 223.

IFSTA, *Essentials of Fire Fighting and Fire Department Operations*, 5th Edition, 1st Printing, page 247.

Jones and Bartlett, NFPA, *Fundamentals of Fire Fighter Skills*, 2nd Edition, 1st Printing, page 184.

Answer: A

131. Reference: NFPA 1001, 5.3.16 and 5.3.16(A)(B)

Delmar, *Firefighter's Handbook*, 3rd Edition, 1st Printing, page 224.

IFSTA, *Essentials of Fire Fighting and Fire Department Operations*, 5th Edition, 1st Printing, page 247.

Jones and Bartlett, NFPA, *Fundamentals of Fire Fighter Skills*, 2nd Edition, 1st Printing, page 184.

Answer: B

132. Reference: NFPA 1001, 5.3.16 and 5.3.16(A)(B)

Delmar, *Firefighter's Handbook*, 3rd Edition, 1st Printing, page 219.

IFSTA, *Essentials of Fire Fighting and Fire Department Operations*, 5th Edition, 1st Printing, page 249.

Jones and Bartlett, NFPA, *Fundamentals of Fire Fighter Skills*, 2nd Edition, 1st Printing, page 184.

Answer: D

133. Reference: NFPA 1001, 5.3.16 and 5.3.16(A)(B)

Delmar, *Firefighter's Handbook*, 3rd Edition, 1st Printing, page 225.

IFSTA, *Essentials of Fire Fighting and Fire Department Operations*, 5th Edition, 1st Printing, page 241.

Jones and Bartlett, NFPA, *Fundamentals of Fire Fighter Skills*, 2nd Edition, 1st Printing, page 198.

Answer: A

134. Reference: NFPA 1001, 5.3.16 and 5.3.16(A)

Delmar, *Firefighter's Handbook*, 3rd Edition, 1st Printing, page 216.

IFSTA, *Essentials of Fire Fighting and Fire Department Operations*, 5th Edition, 1st Printing, page 248.

Jones and Bartlett, NFPA, *Fundamentals of Fire Fighter Skills*, 2nd Edition, 1st Printing, page 183.

Answer: D

135. Reference: NFPA 1001, 5.3.16 and 5.3.16(A)

Delmar, *Firefighter's Handbook*, 3rd Edition, 1st Printing, page 217.

IFSTA, *Essentials of Fire Fighting and Fire Department Operations*, 5th Edition, 1st Printing, page 237.

Jones and Bartlett, NFPA, *Fundamentals of Fire Fighter Skills*, 2nd Edition, 1st Printing, page 188.

Answer: B

136. Reference: NFPA 1001, 5.3.16 and 5.3.16(A)(B)

Delmar, *Firefighter's Handbook*, 3rd Edition, 1st Printing, page 218.

IFSTA, *Essentials of Fire Fighting and Fire Department Operations*, 5th Edition, 1st Printing, page 243.

Jones and Bartlett, NFPA, *Fundamentals of Fire Fighter Skills*, 2nd Edition, 1st Printing, page 188.

Answer: D

137. Reference: NFPA 1001, 5.3.16 and 5.3.16(A)(B)

Delmar, *Firefighter's Handbook*, 3rd Edition, 1st Printing, page 225.

IFSTA, *Essentials of Fire Fighting and Fire Department Operations*, 5th Edition, 1st Printing, page 251.

Jones and Bartlett, NFPA, *Fundamentals of Fire Fighter Skills*, 2nd Edition, 1st Printing, page 198.

Answer: A

138. Reference: NFPA 1001, 5.3.16 and 5.3.16(A)(B)

Delmar, *Firefighter's Handbook*, 3rd Edition, 1st Printing, page 341.

IFSTA, *Essentials of Fire Fighting and Fire Department Operations*, 5th Edition, page 98.

Jones and Bartlett, NFPA, *Fundamentals of Fire Fighter Skills*, 2nd Edition, 1st Printing, page 190.

Answer: B

139. Reference: NFPA 1001, 5.3.18 and 5.3.18(A)(B)

Delmar, *Firefighter's Handbook*, 3rd Edition, 1st Printing, page 573.

IFSTA, *Essentials of Fire Fighting and Fire Department Operations*, 5th Edition, 1st Printing, page 369.

Jones and Bartlett, NFPA, *Fundamentals of Fire Fighter Skills*, 2nd Edition, 1st Printing, page 33.

Answer: A

140. Reference: NFPA 1001, 5.3.19 and 5.3.19(A)(B)

Delmar, *Firefighter's Handbook*, 3rd Edition, 1st Printing, pages 685–686.

IFSTA, *Essentials of Fire Fighting and Fire Department Operations*, 5th Edition, 1st Printing, page 803.

Jones and Bartlett, NFPA, *Fundamentals of Fire Fighter Skills*, 2nd Edition, 1st Printing, page 604.

Answer: C

141. Reference: NFPA 1001, 5.5.1 and 5.5.1(A)(B)

Delmar, *Firefighter's Handbook*, 3rd Edition, 1st Printing, page 201.

IFSTA, *Essentials of Fire Fighting and Fire Department Operations*, 5th Edition, 1st Printing, page 206.

Jones and Bartlett, NFPA, *Fundamentals of Fire Fighter Skills*, 2nd Edition, 1st Printing, page 64.

Answer: C

142. Reference: NFPA 1001, 5.5.1 and 5.5.1(A)(B)

Delmar, *Firefighter's Handbook*, 3rd Edition, 1st Printing, page 174.

IFSTA, *Essentials of Fire Fighting and Fire Department Operations*, 5th Edition, 1st Printing, page 206.

Jones and Bartlett, NFPA, *Fundamentals of Fire Fighter Skills*, 2nd Edition, 1st Printing, page 65.

Answer: B

143. Reference: NFPA 1001, 5.5.1 and 5.5.1(A)(B)

Delmar, *Firefighter's Handbook*, 3rd Edition, 1st Printing, page 173.

IFSTA, *Essentials of Fire Fighting and Fire Department Operations*, 5th Edition, 1st Printing, page 206.

Jones and Bartlett, NFPA, *Fundamentals of Fire Fighter Skills*, 2nd Edition, 1st Printing, page 65.

Answer: D

144. Reference: NFPA 1001, 5.5.1 and 5.5.1(A)

Delmar, *Firefighter's Handbook*, 3rd Edition, 1st Printing, page 496.

IFSTA, *Essentials of Fire Fighting and Fire Department Operations*, 5th Edition, 1st Printing, page 269.

Jones and Bartlett, NFPA, *Fundamentals of Fire Fighter Skills*, 2nd Edition, 1st Printing, page 246.

Answer: B

145. Reference: NFPA 1001, 5.5.1 and 5.5.1(A)(B)

Delmar, *Firefighter's Handbook*, 3rd Edition, 1st Printing, pages 422 and 425.

IFSTA, *Essentials of Fire Fighting and Fire Department Operations*, 5th Edition, 1st Printing, page 479.

Jones and Bartlett, NFPA, *Fundamentals of Fire Fighter Skills*, 2nd Edition, 1st Printing, page 333.

Answer: B

146. Reference: NFPA 1001, 5.5.1 and 5.5.1(A)(B)

Delmar, *Firefighter's Handbook*, 3rd Edition, 1st Printing, pages 422 and 425.

IFSTA, *Essentials of Fire Fighting and Fire Department Operations*, 5th Edition, 1st Printing, page 479.

Jones and Bartlett, NFPA, *Fundamentals of Fire Fighter Skills*, 2nd Edition, 1st Printing, pages 332–333.

Answer: D

147. Reference: NFPA 1001, 5.5.1 and 5.5.1(A)
Delmar, *Firefighter's Handbook*, 3rd Edition, 1st Printing, page 424.
IFSTA, *Essentials of Fire Fighting and Fire Department Operations*, 5th Edition, 1st Printing, page 479.
Jones and Bartlett, NFPA, *Fundamentals of Fire Fighter Skills*, 2nd Edition, 1st Printing, page 333.
Answer: C

148. Reference: NFPA 1001, 5.5.2, 5.5.2(A)(B), 5.3.15 and 5.3.15(A)(B)
Delmar, *Firefighter's Handbook*, 3rd Edition, 1st Printing, page 273.
IFSTA, *Essentials of Fire Fighting and Fire Department Operations*, 5th Edition, 1st Printing, page 656.
Jones and Bartlett, NFPA, *Fundamentals of Fire Fighter Skills*, 2nd Edition, 1st Printing, page 493.
Answer: D

149. Reference: NFPA 1001, 5.5.2 and 5.5.2(A)(B)
Delmar, *Firefighter's Handbook*, 3rd Edition, 1st Printing, page 266.
IFSTA, *Essentials of Fire Fighting and Fire Department Operations*, 5th Edition, 1st Printing, page 653.
Jones and Bartlett, NFPA, *Fundamentals of Fire Fighter Skills*, 2nd Edition, 1st Printing, pages 482–483.
Answer: B

150. Reference: NFPA 1001, 5.5.2 and 5.5.2(A)(B)
Delmar, *Firefighter's Handbook*, 3rd Edition, 1st Printing, page 279.
IFSTA, *Essentials of Fire Fighting and Fire Department Operations*, 5th Edition, 1st Printing, pages 662–663.
Jones and Bartlett, NFPA, *Fundamentals of Fire Fighter Skills*, 2nd Edition, 1st Printing, page 499.
Answer: C

Don't forget to enter the information on your Personal Progress Plotter and answer the Yes and No question at the end of the Examination. This step is extremely important for the successful completion of the Systematic Approach to Examination Preparation!

Examination I-2 Answer Key

Directions

Follow these steps carefully for completing the feedback part of SAEP:

1. After calculating your score, look up the answers for the examination items you missed as well as those on which you guessed, even if you guessed correctly. If you are guessing, it means the answer is not perfectly clear. In this process, we are committed to making you as knowledgeable as possible.

2. Enter the number of missed and guessed examination items in the blanks on your Personal Progress Plotter.

3. Highlight the answer in the reference materials. Read the paragraph preceding and the paragraph following the one in which the correct answer is located. Enter the paragraph number and page number next to the guessed or missed examination item on your examination. Count any part of a paragraph at the beginning of the page as one paragraph until you reach the paragraph containing your highlighted answer. This step will help you locate and review your missed and guessed examination items later in the process. This step is essential to learning the material in context and by association. These learning techniques (context/association) are the very backbone of the SAEP approach.

4. Once you have completed the feedback part, you may proceed to the next examination.

1. Reference: NFPA 1001, 5.1.1
   Delmar, *Firefighter's Handbook*, 3rd Edition, 1st Printing, page 694.
   IFSTA, *Essentials of Fire Fighting and Fire Department Operations*, 5th Edition, 1st Printing, page 18.
   Jones and Bartlett, NFPA, *Fundamentals of Fire Fighter Skills*, 2nd Edition, 1st Printing, page 410.
   Answer: C

2. Reference: NFPA 1001, 5.1.1, 5.2.1, and 5.2.1(A)(B)
   Delmar, *Firefighter's Handbook*, 3rd Edition, 1st Printing, page 38.
   IFSTA, *Essentials of Fire Fighting and Fire Department Operations*, 5th Edition, 1st Printing, page 33.
   Jones and Bartlett, NFPA, *Fundamentals of Fire Fighter Skills*, 2nd Edition, 1st Printing, page 6.
   Answer: A

3. Reference: NFPA 1001, 5.1.1, 5.2.1, 5.2.1(A)(B), and 5.3
   Delmar, *Firefighter's Handbook*, 3rd Edition, 1st Printing, page 694.
   IFSTA, *Essentials of Fire Fighting and Fire Department Operations*, 5th Edition, 1st Printing, page 34.
   Jones and Bartlett, NFPA, *Fundamentals of Fire Fighter Skills*, 2nd Edition, 1st Printing, page 410.
   Answer: D

4. Reference: NFPA 1001, 5.1.1

Delmar, *Firefighter's Handbook*, 3rd Edition, 1st Printing, page 43.

IFSTA, *Essentials of Fire Fighting and Fire Department Operations*, 5th Edition, 1st Printing, page 37.

Jones and Bartlett, NFPA, *Fundamentals of Fire Fighter Skills*, 2nd Edition, 1st Printing, page 111.

Answer: C

5. Reference: NFPA 1001, 5.1.1

Delmar, *Firefighter's Handbook*, 3rd Edition, 1st Printing, page 28.

IFSTA, *Essentials of Fire Fighting and Fire Department Operations*, 5th Edition, 1st Printing, page 21.

Jones and Bartlett, NFPA, *Fundamentals of Fire Fighter Skills*, 1st Edition, 1st Printing, page 9.

Answer: B

6. Reference: NFPA 1001, 5.1.1, 5.3

Delmar, *Firefighter's Handbook*, 3rd Edition, 1st Printing, page 43.

IFSTA, *Essentials of Fire Fighting and Fire Department Operations*, 5th Edition, 1st Printing, page 37.

Jones and Bartlett, NFPA, *Fundamentals of Fire Fighter Skills*, 2nd Edition, 1st Printing, page 111.

Answer: B

7. Reference: NFPA 1001, 5.1.1

Delmar, *Firefighter's Handbook*, 3rd Edition, 1st Printing, pages 39 and 49.

IFSTA, *Essentials of Fire Fighting and Fire Department Operations*, 5th Edition, 1st Printing, page 19.

Jones and Bartlett, NFPA, *Fundamentals of Fire Fighter Skills*, 2nd Edition, 1st Printing, page 9.

Answer: B

8. Reference: NFPA 1001, 5.1.1

Delmar, *Firefighter's Handbook*, 3rd Edition, 1st Printing, page 41.

IFSTA, *Essentials of Fire Fighting and Fire Department Operations*, 5th Edition, 1st Printing, page 35.

Jones and Bartlett, NFPA, *Fundamentals of Fire Fighter Skills*, 2nd Edition, 1st Printing, page 107.

Answer: A

9. Reference: NFPA 1001, 5.1.1

Delmar, *Firefighter's Handbook*, 3rd Edition, 1st Printing, page 37.

IFSTA, *Essentials of Fire Fighting and Fire Department Operations*, 5th Edition, 1st Printing, page 33.

Jones and Bartlett, NFPA, *Fundamentals of Fire Fighter Skills*, 2nd Edition, 1st Printing, page 6.

Answer: B

10. Reference: NFPA 1001, 5.1.1

Delmar, *Firefighter's Handbook*, 3rd Edition, 1st Printing, page 39.

IFSTA, *Essentials of Fire Fighting and Fire Department Operations*, 5th Edition, 1st Printing, page 19.

Jones and Bartlett, NFPA, *Fundamentals of Fire Fighter Skills*, 2nd Edition, 1st Printing, page 9.

Answer: C

11. Reference: NFPA 1001, 5.1.1, 5.3.14, and 5.3.14(A)(B)

Delmar, *Firefighter's Handbook*, 3rd Edition, 1st Printing, page 352.

IFSTA, *Essentials of Fire Fighting and Fire Department Operations*, 5th Edition, 1st Printing, page 825.

Jones and Bartlett, NFPA, *Fundamentals of Fire Fighter Skills*, 2nd Edition, 1st Printing, pages 650–651.

Answer: C

12. Reference: NFPA 1001, 5.1.1, 5.3.15, and 5.3.15(A)(B)

Delmar, *Firefighter's Handbook*, 3rd Edition, 1st Printing, pages 243–244.

IFSTA, *Essentials of Fire Fighting and Fire Department Operations*, 5th Edition, 1st Printing, pages 610–612.

Jones and Bartlett, NFPA, *Fundamentals of Fire Fighter Skills*, 2nd Edition, 1st Printing, pages 651 and 444.

Answer: A

13. Reference: NFPA 1001, 5.1.2, 5.3.1, 5.3.1(A), 5.3.2, and 5.3.2(A)(B)

Delmar, *Firefighter's Handbook*, 3rd Edition, 1st Printing, page 141.

IFSTA, *Essentials of Fire Fighting and Fire Department Operations*, 5th Edition, 1st Printing, pages 167–168.

Jones and Bartlett, NFPA, *Fundamentals of Fire Fighter Skills*, 2nd Edition, 1st Printing, page 40.

Answer: B

14. Reference: NFPA 1001, 5.1.1, and 5.1.2

Delmar, *Firefighter's Handbook*, 3rd Edition, 1st Printing, page 481.

IFSTA, *Essentials of Fire Fighting and Fire Department Operations*, 5th Edition, 1st Printing, pages 276.

Jones and Bartlett, NFPA, *Fundamentals of Fire Fighter Skills*, 2nd Edition, 1st Printing, page 251.

Answer: A

15. Reference: NFPA 1001, 5.1.1, and 5.1.2

Delmar, *Firefighter's Handbook*, 3rd Edition, 1st Printing, page 478.

IFSTA, *Essentials of Fire Fighting and Fire Department Operations*, 5th Edition, 1st Printing, page 274.

Jones and Bartlett, NFPA, *Fundamentals of Fire Fighter Skills*, 2nd Edition, 1st Printing, page 248.

Answer: C

16. Reference: NFPA 1001, 5.1.1, and 5.1.2

Delmar, *Firefighter's Handbook*, 3rd Edition, 1st Printing, page 506.

IFSTA, *Essentials of Fire Fighting and Fire Department Operations*, 5th Edition, 1st Printing, page 299.

Jones and Bartlett, NFPA, *Fundamentals of Fire Fighter Skills*, 2nd Edition, 1st Printing, pages 266–268.

Answer: C

17. Reference: NFPA 1001, 5.1.1, and 5.1.2

Delmar, *Firefighter's Handbook*, 3rd Edition, 1st Printing, page 477.

IFSTA, *Essentials of Fire Fighting and Fire Department Operations*, 5th Edition, 1st Printing, page 264.

Jones and Bartlett, NFPA, *Fundamentals of Fire Fighter Skills*, 2nd Edition, 1st Printing, page 238.

Answer: A

18. Reference: NFPA 1001, 5.1.1, and 5.1.2

Delmar, *Firefighter's Handbook*, 3rd Edition, 1st Printing, page 481.

IFSTA, *Essentials of Fire Fighting and Fire Department Operations*, 5th Edition, 1st Printing, page 276.

Jones and Bartlett, NFPA, *Fundamentals of Fire Fighter Skills*, 2nd Edition, 1st Printing, page 251.

Answer: B

19. Reference: NFPA 1001, 5.1.1, and 5.1.2

Delmar, *Firefighter's Handbook*, 3rd Edition, 1st Printing, page 481.

IFSTA, *Essentials of Fire Fighting and Fire Department Operations*, 5th Edition, 1st Printing, page 278.

Jones and Bartlett, NFPA, *Fundamentals of Fire Fighter Skills*, 2nd Edition, 1st Printing, page 260.

Answer: C

20. Reference: NFPA 1001, 5.1.1, and 5.1.2

Delmar, *Firefighter's Handbook*, 3rd Edition, 1st Printing, page 487.

IFSTA, *Essentials of Fire Fighting and Fire Department Operations*, 5th Edition, 1st Printing, page 289.

Jones and Bartlett, NFPA, *Fundamentals of Fire Fighter Skills*, 2nd Edition, 1st Printing, page 261.

Answer: A

21. Reference: NFPA 1001, 5.1.1, and 5.1.2

Delmar, *Firefighter's Handbook*, 3rd Edition, 1st Printing, page 484.

IFSTA, *Essentials of Fire Fighting and Fire Department Operations*, 5th Edition, 1st Printing, page 293.

Jones and Bartlett, NFPA, *Fundamentals of Fire Fighter Skills*, 2nd Edition, 1st Printing, page 262.

Answer: B

22. Reference: NFPA 1001, 5.1.1, and 5.1.2

Delmar, *Firefighter's Handbook*, 3rd Edition, 1st Printing, page 479.

IFSTA, *Essentials of Fire Fighting and Fire Department Operations*, 5th Edition, 1st Printing, page 275.

Jones and Bartlett, NFPA, *Fundamentals of Fire Fighter Skills*, 2nd Edition, 1st Printing, page 249.

Answer: B

23. Reference: NFPA 1001, 5.1.1, and 5.1.2

Delmar, *Firefighter's Handbook*, 3rd Edition, 1st Printing, pages 478–479.

IFSTA, *Essentials of Fire Fighting and Fire Department Operations*, 5th Edition, 1st Printing, pages 274–275.

Jones and Bartlett, NFPA, *Fundamentals of Fire Fighter Skills*, 2nd Edition, 1st Printing, page 248.

Answer: A

24. Reference: NFPA 1001, 5.1.1 and 5.1.2

Delmar, *Firefighter's Handbook*, 3rd Edition, 1st Printing, page 478.

IFSTA, *Essentials of Fire Fighting and Fire Department Operations*, 5th Edition, 1st Printing, page 274.

Jones and Bartlett, NFPA, *Fundamentals of Fire Fighter Skills*, 2nd Edition, 1st Printing, page 248.

Answer: D

25. Reference: NFPA 1001, 5.1.1 and 5.1.2

Delmar, *Firefighter's Handbook*, 3rd Edition, 1st Printing, page 505.

IFSTA, *Essentials of Fire Fighting and Fire Department Operations*, 5th Edition, 1st Printing, pages 281 and 296.

Jones and Bartlett, NFPA, *Fundamentals of Fire Fighter Skills*, 2nd Edition, 1st Printing, page 264.

Answer: C

26. Reference: NFPA 1001, 5.1.1 and 5.1.2

Delmar, *Firefighter's Handbook*, 3rd Edition, 1st Printing, pages 505–506.

IFSTA, *Essentials of Fire Fighting and Fire Department Operations*, 5th Edition, 1st Printing, page 280.

Jones and Bartlett, NFPA, *Fundamentals of Fire Fighter Skills*, 2nd Edition, 1st Printing, page 264.

Answer: D

27. Reference: NFPA 1001, 5.1.1, and 5.1.2

Delmar, *Firefighter's Handbook*, 3rd Edition, 1st Printing, page 479.

IFSTA, *Essentials of Fire Fighting and Fire Department Operations*, 5th Edition, 1st Printing, page 275.

Jones and Bartlett, NFPA, *Fundamentals of Fire Fighter Skills*, 2nd Edition, 1st Printing, page 249.

Answer: C

28. Reference: NFPA 1001, 5.1.1, and 5.1.2

Delmar, *Firefighter's Handbook*, 3rd Edition, 1st Printing, page 479.

IFSTA, *Essentials of Fire Fighting and Fire Department Operations*, 5th Edition, 1st Printing, page 275.

Jones and Bartlett, NFPA, *Fundamentals of Fire Fighter Skills*, 2nd Edition, 1st Printing, page 249.

Answer: A

29. Reference: NFPA 1001, 5.1.1 and 5.1.2

Delmar, *Firefighter's Handbook*, 3rd Edition, 1st Printing, page 473.

IFSTA, *Essentials of Fire Fighting and Fire Department Operations*, 5th Edition, 1st Printing, pages 265 and 266.

Jones and Bartlett, NFPA, *Fundamentals of Fire Fighter Skills*, 2nd Edition, 1st Printing, page 240.

Answer: B

30. Reference: NFPA 1001, 5.1.1, and 5.1.2

Delmar, *Firefighter's Handbook*, 3rd Edition, 1st Printing, page 479.

IFSTA, *Essentials of Fire Fighting and Fire Department Operations*, 5th Edition, 1st Printing, page 275.

Jones and Bartlett, NFPA, *Fundamentals of Fire Fighter Skills*, 2nd Edition, 1st Printing, page 250.

Answer: C

31. Reference: NFPA 1001, 5.1.1, and 5.1.2

Delmar, *Firefighter's Handbook*, 3rd Edition, 1st Printing, pages 483–484.

IFSTA, *Essentials of Fire Fighting and Fire Department Operations*, 5th Edition, 1st Printing, page 293.

Jones and Bartlett, NFPA, *Fundamentals of Fire Fighter Skills*, 2nd Edition, 1st Printing, page 262.

Answer: B

32. Reference: NFPA 1001, 5.1.1

Delmar, *Firefighter's Handbook*, 3rd Edition, 1st Printing, page 100.

IFSTA, *Essentials of Fire Fighting and Fire Department Operations*, 5th Edition, 1st Printing, page 113.

Jones and Bartlett, NFPA, *Fundamentals of Fire Fighter Skills*, 2nd Edition, 1st Printing, page 135.

Answer: B

33. Reference: NFPA 1001, 5.2.1 and 5.2.1(A)(B)

Delmar, *Firefighter's Handbook*, 3rd Edition, 1st Printing, page 72.

IFSTA, *Essentials of Fire Fighting and Fire Department Operations*, 5th Edition, 1st Printing, page 943.

Jones and Bartlett, NFPA, *Fundamentals of Fire Fighter Skills*, 2nd Edition, 1st Printing, page 91.

Answer: C

34. Reference: NFPA 1001, 5.2.1 and 5.2.1(A)(B)

Delmar, *Firefighter's Handbook*, 3rd Edition, 1st Printing, page 56.

IFSTA, *Essentials of Fire Fighting and Fire Department Operations*, 5th Edition, 1st Printing, page 929.

Jones and Bartlett, NFPA, *Fundamentals of Fire Fighter Skills*, 2nd Edition, 1st Printing, page 80.

Answer: B

35. Reference: NFPA 1001, 5.2.1, 5.2.1(A)(B), 5.2.3, and 5.2.3(A)(B)

Delmar, *Firefighter's Handbook*, 3rd Edition, 1st Printing, page 66.

IFSTA, *Essentials of Fire Fighting and Fire Department Operations*, 5th Edition, 1st Printing, page 927.

Jones and Bartlett, NFPA, *Fundamentals of Fire Fighter Skills*, 2nd Edition, 1st Printing, page 86.

Answer: C

36. Reference: NFPA 1001, 5.2.1 and 5.2.1(A)(B)

Delmar, *Firefighter's Handbook*, 3rd Edition, 1st Printing, pages 58 and 1105.

IFSTA, *Essentials of Fire Fighting and Fire Department Operations*, 5th Edition, 1st Printing, page 934.

Jones and Bartlett, NFPA, *Fundamentals of Fire Fighter Skills*, 2nd Edition, 1st Printing, page 81.

Answer: A

37. Reference: NFPA 1001, 5.2.1 and 5.2.1(A)(B)

Delmar, *Firefighter's Handbook*, 3rd Edition, 1st Printing, page 59.

IFSTA, *Essentials of Fire Fighting and Fire Department Operations*, 5th Edition, 1st Printing, page 929.

Jones and Bartlett, NFPA, *Fundamentals of Fire Fighter Skills*, 2nd Edition, 1st Printing, page 80.

Answer: C

38. Reference: NFPA 1001, 5.2.1 and 5.2.1(A)

Delmar, *Firefighter's Handbook*, 3rd Edition, 1st Printing, page 61.

IFSTA, *Essentials of Fire Fighting and Fire Department Operations*, 5th Edition, 1st Printing, page 938.

Jones and Bartlett, NFPA, *Fundamentals of Fire Fighter Skills*, 2nd Edition, 1st Printing, page 85.

Answer: A

39. Reference: NFPA 1001, 5.2.1 and 5.2.1(A)(B)

Delmar, *Firefighter's Handbook*, 3rd Edition, 1st Printing, page 77.

IFSTA, *Essentials of Fire Fighting and Fire Department Operations*, 5th Edition, 1st Printing, page 935.

Jones and Bartlett, NFPA, *Fundamentals of Fire Fighter Skills*, 2nd Edition, 1st Printing, page 82.

Answer: A

40. Reference: NFPA 1001, 5.2.1 and 5.2.1(A)(B)

Delmar, *Firefighter's Handbook*, 3rd Edition, 1st Printing, page 53.

IFSTA, *Essentials of Fire Fighting and Fire Department Operations*, 5th Edition, 1st Printing, page 936.

Jones and Bartlett, NFPA, *Fundamentals of Fire Fighter Skills*, 2nd Edition, 1st Printing, page 82.

Answer: C

41. Reference: NFPA 1001, 5.2.1 and 5.2.1(A)(B)

Delmar, *Firefighter's Handbook*, 3rd Edition, 1st Printing, page 60.

IFSTA, *Essentials of Fire Fighting and Fire Department Operations*, 5th Edition, 1st Printing, page 936.

Jones and Bartlett, NFPA, *Fundamentals of Fire Fighter Skills*, 2nd Edition, 1st Printing, page 83.

Answer: D

42. Reference: NFPA 1001, 5.2.3 and 5.2.3(A)(B)

Delmar, *Firefighter's Handbook*, 3rd Edition, 1st Printing, pages 73–74.

IFSTA, *Essentials of Fire Fighting and Fire Department Operations*, 5th Edition, 1st Printing, page 945.

Jones and Bartlett, NFPA, *Fundamentals of Fire Fighter Skills*, 2nd Edition, 1st Printing, page 93.

Answer: C

43. Reference: NFPA 1001, 5.2.3 and 5.2.3(A)(B)

Delmar, *Firefighter's Handbook*, 3rd Edition, 1st Printing, page 77.

IFSTA, *Essentials of Fire Fighting and Fire Department Operations*, 5th Edition, 1st Printing, page 320.

Jones and Bartlett, NFPA, *Fundamentals of Fire Fighter Skills*, 2nd Edition, 1st Printing, page 96.

Answer: B

44. Reference: NFPA 1001, 5.2.3 and 5.2.3(A)(B)

Delmar, *Firefighter's Handbook*, 3rd Edition, 1st Printing, page 73.

IFSTA, *Essentials of Fire Fighting and Fire Department Operations*, 5th Edition, 1st Printing, page 945.

Jones and Bartlett, NFPA, *Fundamentals of Fire Fighter Skills*, 2nd Edition, 1st Printing, page 93.

Answer: D

45. Reference: NFPA 1001, 5.3.1 and 5.3.1(A)(B)

Delmar, *Firefighter's Handbook*, 3rd Edition, 1st Printing, page 202.

IFSTA, *Essentials of Fire Fighting and Fire Department Operations*, 5th Edition, 1st Printing, page 229.

Jones and Bartlett, NFPA, *Fundamentals of Fire Fighter Skills*, 2nd Edition, 1st Printing, page 93.

Answer: B

46. Reference: NFPA 1001, 5.3.1, 5.3.1(A)(B), 5.3.5, 5.3.5(A)(B), 5.3.9, and 5.3.9(A)(B)
Delmar, *Firefighter's Handbook*, 3rd Edition, 1st Printing, page 177.
IFSTA, *Essentials of Fire Fighting and Fire Department Operations*, 5th Edition, 1st Printing, page 208.
Jones and Bartlett, NFPA, *Fundamentals of Fire Fighter Skills*, 2nd Edition, 1st Printing, page 52.
Answer: B

47. Reference: NFPA 1001, 5.3.1, 5.3.1(A), 5.3.5, 5.3.5(A)(B), 5.3.10, 5.3.10(A), 5.3.11, and 5.3.11(A)
Delmar, *Firefighter's Handbook*, 3rd Edition, 1st Printing, page 162.
IFSTA, *Essentials of Fire Fighting and Fire Department Operations*, 5th Edition, 1st Printing, page 180.
Jones and Bartlett, NFPA, *Fundamentals of Fire Fighter Skills*, 2nd Edition, 1st Printing, pages 47–48.
Answer: D

48. Reference: NFPA 1001, 5.3.1, 5.3.1(A)(B), and 5.1.2
Delmar, *Firefighter's Handbook*, 3rd Edition, 1st Printing, pages 187–188.
IFSTA, *Essentials of Fire Fighting and Fire Department Operations*, 5th Edition, 1st Printing, page 203.
Jones and Bartlett, NFPA, *Fundamentals of Fire Fighter Skills*, 2nd Edition, 1st Printing, page 57.
Answer: A

49. Reference: NFPA 1001, 5.3.1, 5.3.1(A)(B), 5.3.5, and 5.3.5(A)(B)
Delmar, *Firefighter's Handbook*, 3rd Edition, 1st Printing, page 162.
IFSTA, *Essentials of Fire Fighting and Fire Department Operations*, 5th Edition, 1st Printing, page 180.
Jones and Bartlett, NFPA, *Fundamentals of Fire Fighter Skills*, 2nd Edition, 1st Printing, pages 47–48.
Answer: A

50. Reference: NFPA 1001, 5.3.1, 5.3.1(A), 5.3.5, 5.3.5(A)(B), 5.3.9, and 5.3.9(A)(B)
Delmar, *Firefighter's Handbook*, 3rd Edition, 1st Printing, page 573.
IFSTA, *Essentials of Fire Fighting and Fire Department Operations*, 5th Edition, 1st Printing, page 186.
Jones and Bartlett, NFPA, *Fundamentals of Fire Fighter Skills*, 2nd Edition, 1st Printing, page 48.
Answer: D

51. Reference: NFPA 1001, 5.3.1, 5.3.1(A)(B), 5.3.5, 5.3.5(A)(B), 5.3.9, 5.3.9(A)(B), 5.3.10, and 5.3.10(A)
Delmar, *Firefighter's Handbook*, 3rd Edition, 1st Printing, pages 161–162.
IFSTA, *Essentials of Fire Fighting and Fire Department Operations*, 5th Edition, 1st Printing, page 198.
Jones and Bartlett, NFPA, *Fundamentals of Fire Fighter Skills*, 2nd Edition, 1st Printing, pages 40 and 536.
Answer: B

52. Reference: NFPA 1001, 5.3.1, 5.3.1(A), 5.3.5, and 5.3.5(A)(B)

Delmar, *Firefighter's Handbook*, 3rd Edition, 1st Printing, page 178.

IFSTA, *Essentials of Fire Fighting and Fire Department Operations*, 5th Edition, 1st Printing, page 198.

Jones and Bartlett, NFPA, *Fundamentals of Fire Fighter Skills*, 2nd Edition, 1st Printing, page 49.

Answer: B

53. Reference: NFPA 1001, 5.3.1, 5.3.1(A)(B), and 5.3.5(A)(B)

Delmar, *Firefighter's Handbook*, 3rd Edition, 1st Printing, page 172.

IFSTA, *Essentials of Fire Fighting and Fire Department Operations*, 5th Edition, 1st Printing, page 190.

Jones and Bartlett, NFPA, *Fundamentals of Fire Fighter Skills*, 2nd Edition, 1st Printing, page 49.

Answer: B

54. Reference: NFPA 1001, 5.3.1, 5.3.1(A)(B), 5.3.5, and 5.3.5(A)(B)

Delmar, *Firefighter's Handbook*, 3rd Edition, 1st Printing, page 178.

IFSTA, *Essentials of Fire Fighting and Fire Department Operations*, 5th Edition, 1st Printing, page 190.

Jones and Bartlett, NFPA, *Fundamentals of Fire Fighter Skills*, 2nd Edition, 1st Printing, page 49.

Answer: D

55. Reference: NFPA 1001, 5.3.1 and 5.3.1(A)(B)

Delmar, *Firefighter's Handbook*, 3rd Edition, 1st Printing, pages 177 and 186.

IFSTA, *Essentials of Fire Fighting and Fire Department Operations*, 5th Edition, 1st Printing, page 203.

Jones and Bartlett, NFPA, *Fundamentals of Fire Fighter Skills*, 2nd Edition, 1st Printing, page 53.

Answer: B

56. Reference: NFPA 1001, 5.3.2, 5.3.2(A)(B), 5.1.1, 5.3.3, and 5.3.3(A)(B)

Delmar, *Firefighter's Handbook*, 3rd Edition, page 143.

IFSTA, *Essentials of Fire Fighting and Fire Department Operations*, 5th Edition, page 168.

Jones and Bartlett, NFPA, *Fundamentals of Fire Fighter Skills*, 2nd Edition, 1st Printing, page 39.

Answer: C

57. Reference: NFPA 1001, 5.3.2 and 5.3.2(A)(B)

Delmar, *Firefighter's Handbook*, 3rd Edition, 1st Printing, page 141.

IFSTA, *Essentials of Fire Fighting and Fire Department Operations*, 5th Edition, 1st Printing, page 165.

Jones and Bartlett, NFPA, *Fundamentals of Fire Fighter Skills*, 2nd Edition, 1st Printing, page 35.

Answer: B

58. Reference: NFPA 1001, 5.3.2 and 5.3.2(A)(B)

Delmar, *Firefighter's Handbook*, 3rd Edition, 1st Printing, page 143.

IFSTA, *Essentials of Fire Fighting and Fire Department Operations*, 5th Edition, 1st Printing, page 176.

Jones and Bartlett, NFPA, *Fundamentals of Fire Fighter Skills*, 2nd Edition, 1st Printing, page 39.

Answer: D

59. Reference: NFPA 1001, 5.3.3 and 5.3.3(A)

Delmar, *Firefighter's Handbook*, 3rd Edition, 1st Printing, pages 829–830.

IFSTA, *Essentials of Fire Fighting and Fire Department Operations*, 5th Edition, page 70.

Jones and Bartlett, NFPA, *Fundamentals of Fire Fighter Skills*, 2nd Edition, 1st Printing, page 285.

Answer: B

60. Reference: NFPA 1001, 5.3.4 and 5.3.4(A)(B)

Delmar, *Firefighter's Handbook*, 3rd Edition, 1st Printing, page 591.

IFSTA, *Essentials of Fire Fighting and Fire Department Operations*, 5th Edition, 1st Printing, page 398.

Jones and Bartlett, NFPA, *Fundamentals of Fire Fighter Skills*, 2nd Edition, 1st Printing, page 221.

Answer: A

61. Reference: NFPA 1001, 5.3.4 and 5.3.4(A)(B)

Delmar, *Firefighter's Handbook*, 3rd Edition, 1st Printing, page 594.

IFSTA, *Essentials of Fire Fighting and Fire Department Operations*, 5th Edition, 1st Printing, page 434.

Jones and Bartlett, NFPA, *Fundamentals of Fire Fighter Skills*, 2nd Edition, 1st Printing, page 313.

Answer: A

62. Reference: NFPA 1001, 5.3.4 and 5.3.4(A)

Delmar, *Firefighter's Handbook*, 3rd Edition, 1st Printing, page 598.

IFSTA, *Essentials of Fire Fighting and Fire Department Operations*, 5th Edition, 1st Printing, page 415.

Jones and Bartlett, NFPA, *Fundamentals of Fire Fighter Skills*, 2nd Edition, 1st Printing, page 298.

Answer: A

63. Reference: NFPA 1001, 5.3.4 and 5.3.4(A)(B)

Delmar, *Firefighter's Handbook*, 3rd Edition, 1st Printing, pages 594–595.

IFSTA, *Essentials of Fire Fighting and Fire Department Operations*, 5th Edition, 1st Printing, page 435.

Jones and Bartlett, NFPA, *Fundamentals of Fire Fighter Skills*, 2nd Edition, 1st Printing, page 313.

Answer: B

64. Reference: NFPA 1001, 5.3.4 and 5.3.4(A)(B)

Delmar, *Firefighter's Handbook*, 3rd Edition, 1st Printing, page 601.

IFSTA, *Essentials of Fire Fighting and Fire Department Operations*, 5th Edition, 1st Printing, page 418.

Jones and Bartlett, NFPA, *Fundamentals of Fire Fighter Skills*, 2nd Edition, 1st Printing, page 302.

Answer: D

65. Reference: NFPA 1001, 5.3.4 and 5.3.4(A)

Delmar, *Firefighter's Handbook*, 3rd Edition, 1st Printing, page 598.

IFSTA, *Essentials of Fire Fighting and Fire Department Operations*, 5th Edition, 1st Printing, page 414.

Jones and Bartlett, NFPA, *Fundamentals of Fire Fighter Skills*, 2nd Edition, 1st Printing, page 297.

Answer: A

66. Reference: NFPA 1001, 5.3.4 and 5.3.4(A)(B)

Delmar, *Firefighter's Handbook*, 3rd Edition, 1st Printing, page 590.

IFSTA, *Essentials of Fire Fighting and Fire Department Operations*, 5th Edition, 1st Printing, page 404.

Jones and Bartlett, NFPA, *Fundamentals of Fire Fighter Skills*, 2nd Edition, 1st Printing, page 219.

Answer: D

67. Reference: NFPA 1001, 5.3.4 and 5.3.4(A)(B)

Delmar, *Firefighter's Handbook*, 3rd Edition, 1st Printing, page 620.

IFSTA, *Essentials of Fire Fighting and Fire Department Operations*, 5th Edition, 1st Printing, page 439.

Jones and Bartlett, NFPA, *Fundamentals of Fire Fighter Skills*, 2nd Edition, 1st Printing, page 303.

Answer: D

68. Reference: NFPA 1001, 5.3.5, 5.3.5(A)(B), 5.3.9, and 5.3.9(A)(B)

Delmar, *Firefighter's Handbook*, 3rd Edition, 1st Printing, page 520.

IFSTA, *Essentials of Fire Fighting and Fire Department Operations*, 5th Edition, 1st Printing, page 309.

Jones and Bartlett, NFPA, *Fundamentals of Fire Fighter Skills*, 2nd Edition, 1st Printing, page 369.

Answer: B

69. Reference: NFPA 1001, 5.3.5, 5.3.5(A)(B), 5.3.9, and 5.3.9(A)(B)

Delmar, *Firefighter's Handbook*, 3rd Edition, 1st Printing, page 520.

IFSTA, *Essentials of Fire Fighting and Fire Department Operations*, 5th Edition, 1st Printing, page 309.

Jones and Bartlett, NFPA, *Fundamentals of Fire Fighter Skills*, 2nd Edition, 1st Printing, page 534.

Answer: C

70. Reference: NFPA 1001, 5.3.5, 5.3.5(A)(B), 5.3.9, and 5.3.9(A)(B)

Delmar, *Firefighter's Handbook*, 3rd Edition, 1st Printing, page 520.

IFSTA, *Essentials of Fire Fighting and Fire Department Operations*, 5th Edition, 1st Printing, pages 308–309.

Jones and Bartlett, NFPA, *Fundamentals of Fire Fighter Skills*, 2nd Edition, 1st Printing, page 367.

Answer: D

71. Reference: NFPA 1001, 5.3.5 and 5.3.5(A)

Delmar, *Firefighter's Handbook*, 3rd Edition, 1st Printing, page 825.

IFSTA, *Essentials of Fire Fighting and Fire Department Operations*, 5th Edition, 1st Printing, pages 74 and 75.

Jones and Bartlett, NFPA, *Fundamentals of Fire Fighter Skills*, 2nd Edition, 1st Printing, page 32.

Answer: B

72. Reference: NFPA 1001, 5.3.5 and 5.3.5(A)(B)

Delmar, *Firefighter's Handbook*, 3rd Edition, 1st Printing, page 523.

IFSTA, *Essentials of Fire Fighting and Fire Department Operations*, 5th Edition, 1st Printing, page 77.

Jones and Bartlett, NFPA, *Fundamentals of Fire Fighter Skills*, 2nd Edition, 1st Printing, page 538.

Answer: A

73. Reference: NFPA 1001, 5.3.5 and 5.3.5(A)

Delmar, *Firefighter's Handbook*, 3rd Edition, 1st Printing, pages 824–825.

IFSTA, *Essentials of Fire Fighting and Fire Department Operations*, 5th Edition, 1st Printing, page 74.

Jones and Bartlett, NFPA, *Fundamentals of Fire Fighter Skills*, 2nd Edition, 1st Printing, pages 32 and 534.

Answer: A

74. Reference: NFPA 1001, 5.3.5 and 5.3.5(A)

Delmar, *Firefighter's Handbook*, 3rd Edition, 1st Printing, page 825.

IFSTA, *Essentials of Fire Fighting and Fire Department Operations*, 5th Edition, 1st Printing, page 74.

Jones and Bartlett, NFPA, *Fundamentals of Fire Fighter Skills*, 2nd Edition, 1st Printing, page 33.

Answer: B

75. Reference: NFPA 1001, 5.3.9, 5.3.9(A(B), 5.3.10, and 5.3.10(A)(B)

Delmar, *Firefighter's Handbook*, 3rd Edition, 1st Printing, pages 189–190.

IFSTA, *Essentials of Fire Fighting and Fire Department Operations*, 5th Edition, 1st Printing, page 310.

Jones and Bartlett, NFPA, *Fundamentals of Fire Fighter Skills*, 2nd Edition, 1st Printing, page 374.

Answer: C

76. Reference: NFPA 1001, 5.3.9 and 5.3.9(A)

Delmar, *Firefighter's Handbook*, 3rd Edition, 1st Printing, page 519.

IFSTA, *Essentials of Fire Fighting and Fire Department Operations*, 5th Edition, 1st Printing, page 305.

Jones and Bartlett, NFPA, *Fundamentals of Fire Fighter Skills*, 2nd Edition, 1st Printing, page 366.

Answer: A

77. Reference: NFPA 1001, 5.3.9 and 5.3.9(A)(B)

Delmar, *Firefighter's Handbook*, 3rd Edition, 1st Printing, page 533.

IFSTA, *Essentials of Fire Fighting and Fire Department Operations*, 5th Edition, 1st Printing, page 328.

Jones and Bartlett, NFPA, *Fundamentals of Fire Fighter Skills*, 2nd Edition, 1st Printing, page 382.

Answer: B

78. Reference: NFPA 1001, 5.3.9 and 5.3.9(A)(B)

Delmar, *Firefighter's Handbook*, 3rd Edition, 1st Printing, page 535.

IFSTA, *Essentials of Fire Fighting and Fire Department Operations*, 5th Edition, 1st Printing, page 330.

Jones and Bartlett, NFPA, *Fundamentals of Fire Fighter Skills*, 2nd Edition, 1st Printing, pages 380 and 383.

Answer: C

79. Reference: NFPA 1001, 5.3.9 and 5.3.9(A)(B)

Delmar, *Firefighter's Handbook*, 3rd Edition, 1st Printing, page 536.

IFSTA, *Essentials of Fire Fighting and Fire Department Operations*, 5th Edition, 1st Printing, page 328.

Jones and Bartlett, NFPA, *Fundamentals of Fire Fighter Skills*, 2nd Edition, 1st Printing, page 385.

Answer: B

80. Reference: NFPA 1001, 5.3.9 and 5.3.9(A)(B)

Delmar, *Firefighter's Handbook*, 3rd Edition, 1st Printing, page 538.

IFSTA, *Essentials of Fire Fighting and Fire Department Operations*, 5th Edition, 1st Printing, page 328.

Jones and Bartlett, NFPA, *Fundamentals of Fire Fighter Skills*, 2nd Edition, 1st Printing, page 379.

Answer: B

81. Reference: NFPA 1001, 5.3.9 and 5.3.9(A)

Delmar, *Firefighter's Handbook*, 3rd Edition, 1st Printing, page 525.

IFSTA, *Essentials of Fire Fighting and Fire Department Operations*, 5th Edition, 1st Printing, page 310.

Jones and Bartlett, NFPA, *Fundamentals of Fire Fighter Skills*, 2nd Edition, 1st Printing, page 369.

Answer: A

82. Reference: NFPA 1001, 5.3.10 and 5.3.10(A)(B)
Delmar, *Firefighter's Handbook*, 3rd Edition, 1st Printing, page 261.
IFSTA, *Essentials of Fire Fighting and Fire Department Operations*, 5th Edition, 1st Printing, page 649.
Jones and Bartlett, NFPA, *Fundamentals of Fire Fighter Skills*, 2nd Edition, 1st Printing, page 481.
Answer: D

83. Reference: NFPA 1001, 5.3.10 and 5.3.10(A)(B)
Delmar, *Firefighter's Handbook*, 3rd Edition, 1st Printing, page 326.
IFSTA, *Essentials of Fire Fighting and Fire Department Operations*, 5th Edition, 1st Printing, page 726.
Jones and Bartlett, NFPA, *Fundamentals of Fire Fighter Skills*, 2nd Edition, 1st Printing, page 517.
Answer: A

84. Reference: NFPA 1001, 5.3.10 and 5.3.10(A)(B)
Delmar, *Firefighter's Handbook*, 3rd Edition, 1st Printing, pages 326 and 328.
IFSTA, *Essentials of Fire Fighting and Fire Department Operations*, 5th Edition, 1st Printing, page 768.
Jones and Bartlett, NFPA, *Fundamentals of Fire Fighter Skills*, 2nd Edition, 1st Printing, page 621.
Answer: A

85. Reference: NFPA 1001, 5.3.10, 5.3.10(A)(B), 5.3.8 and 5.3.8(A)(B)
Delmar, *Firefighter's Handbook*, 3rd Edition, 1st Printing, pages 335 and 1109.
IFSTA, *Essentials of Fire Fighting and Fire Department Operations*, 5th Edition, 1st Printing, page 722.
Jones and Bartlett, NFPA, *Fundamentals of Fire Fighter Skills*, 2nd Edition, 1st Printing, page 465.
Answer: B

86. Reference: NFPA 1001, 5.3.10, 5.3.10(A)(B), 5.3.8 and 5.3.8(A)(B)
Delmar, *Firefighter's Handbook*, 3rd Edition, 2nd Printing, page 260.
IFSTA, *Essentials of Fire Fighting and Fire Department Operations*, 5th Edition, 1st Printing, page 643.
Jones and Bartlett, NFPA, *Fundamentals of Fire Fighter Skills*, 2nd Edition, 1st Printing, page 479.
Answer: B

87. Reference: NFPA 1001, 5.3.10 and 5.3.10(A)(B)
Delmar, *Firefighter's Handbook*, 3rd Edition, 1st Printing, page 373.
IFSTA, *Essentials of Fire Fighting and Fire Department Operations*, 5th Edition, 1st Printing, page 674.
Jones and Bartlett, NFPA, *Fundamentals of Fire Fighter Skills*, 2nd Edition, 1st Printing, page 514.
Answer: D

88. Reference: NFPA 1001, 5.3.10, 5.3.10(A)(B), 5.3.8 and 5.3.8(A)(B)
Delmar, *Firefighter's Handbook*, 3rd Edition, 1st Printing, page 273.
IFSTA, *Essentials of Fire Fighting and Fire Department Operations*, 5th Edition, 1st Printing, page 658.
Jones and Bartlett, NFPA, *Fundamentals of Fire Fighter Skills*, 2nd Edition, 1st Printing, page 495.
Answer: D

89. Reference: NFPA 1001, 5.3.10, 5.3.10(A)(B), 5.3.8 and 5.3.8(A)(B)
Delmar, *Firefighter's Handbook*, 3rd Edition, 1st Printing, page 287.
IFSTA, *Essentials of Fire Fighting and Fire Department Operations*, 5th Edition, 1st Printing, page 789.
Jones and Bartlett, NFPA, *Fundamentals of Fire Fighter Skills*, 2nd Edition, 1st Printing, page 6.
Answer: C

90. Reference: NFPA 1001, 5.3.10, 5.3.10(A)(B), 5.3.8 and 5.3.8(A)(B)
Delmar, *Firefighter's Handbook*, 3rd Edition, 1st Printing, page 296.
IFSTA, *Essentials of Fire Fighting and Fire Department Operations*, 5th Edition, 1st Printing, page 675.
Jones and Bartlett, NFPA, *Fundamentals of Fire Fighter Skills*, 2nd Edition, 1st Printing, pages 510–511.
Answer: A

91. Reference: NFPA 1001, 5.3.10 and 5.3.10(A)(B)
Delmar, *Firefighter's Handbook*, 3rd Edition, 1st Printing, pages 328–329.
IFSTA, *Essentials of Fire Fighting and Fire Department Operations*, 5th Edition, 1st Printing, page 731.
Jones and Bartlett, NFPA, *Fundamentals of Fire Fighter Skills*, 2nd Edition, 1st Printing, page 518.
Answer: D

92. Reference: NFPA 1001, 5.3.10 and 5.3.10(A)
Delmar, *Firefighter's Handbook*, 3rd Edition, page 262.
IFSTA, *Essentials of Fire Fighting and Fire Department Operations*, 5th Edition, page 647.
Jones and Bartlett, NFPA, *Fundamentals of Fire Fighter Skills*, 2nd Edition, 1st Printing, page 480.
Answer: A

93. Reference: NFPA 1001, 5.3.10 and 5.3.10(A)
Delmar, *Firefighter's Handbook*, 3rd Edition, page 262.
IFSTA, *Essentials of Fire Fighting and Fire Department Operations*, 5th Edition, page 647.
Jones and Bartlett, NFPA, *Fundamentals of Fire Fighter Skills*, 2nd Edition, 1st Printing, page 480.
Answer: D

94. Reference: NFPA 1001, 5.3.11, 5.3.11(A), 5.3.12, and 5.3.12(A)

Delmar, *Firefighter's Handbook*, 3rd Edition, 1st Printing, page 631.

IFSTA, *Essentials of Fire Fighting and Fire Department Operations*, 5th Edition, 1st Printing, page 541.

Jones and Bartlett, NFPA, *Fundamentals of Fire Fighter Skills*, 2nd Edition, 1st Printing, page 403.

Answer: D

95. Reference: NFPA 1001, 5.3.11 and 5.3.11(A)

Delmar, *Firefighter's Handbook*, 3rd Edition, 1st Printing, page 645.

IFSTA, *Essentials of Fire Fighting and Fire Department Operations*, 5th Edition, 1st Printing, page 573.

Jones and Bartlett, NFPA, *Fundamentals of Fire Fighter Skills*, 2nd Edition, 1st Printing, page 417.

Answer: A

96. Reference: NFPA 1001, 5.3.11 and 5.3.11(A)(B)

Delmar, *Firefighter's Handbook*, 3rd Edition, 1st Printing, page 644.

IFSTA, *Essentials of Fire Fighting and Fire Department Operations*, 5th Edition, 1st Printing, page 574.

Jones and Bartlett, NFPA, *Fundamentals of Fire Fighter Skills*, 2nd Edition, 1st Printing, page 415.

Answer: B

97. Reference: NFPA 1001, 5.3.11 and 5.3.11(A)(B)

Delmar, *Firefighter's Handbook*, 3rd Edition, 1st Printing, page 644.

IFSTA, *Essentials of Fire Fighting and Fire Department Operations*, 5th Edition, 1st Printing, page 572.

Jones and Bartlett, NFPA, *Fundamentals of Fire Fighter Skills*, 2nd Edition, 1st Printing, page 418.

Answer: C

98. Reference: NFPA 1001, 5.3.11 and 5.3.11(A)(B)

Delmar, *Firefighter's Handbook*, 3rd Edition, 1st Printing, page 669.

IFSTA, *Essentials of Fire Fighting and Fire Department Operations*, 5th Edition, 1st Printing, page 574.

Jones and Bartlett, NFPA, *Fundamentals of Fire Fighter Skills*, 2nd Edition, 1st Printing, page 413.

Answer: C

99. Reference: NFPA 1001, 5.3.11, 5.3.11(A), 5.3.12 and 5.3.12(A)

Delmar, *Firefighter's Handbook*, 3rd Edition, 1st Printing, page 97.

IFSTA, *Essentials of Fire Fighting and Fire Department Operations*, 5th Edition, 1st Printing, page 95.

Jones and Bartlett, NFPA, *Fundamentals of Fire Fighter Skills*, 2nd Edition, 1st Printing, page 132.

Answer: D

100. Reference: NFPA 1001, 5.3.12 and 5.3.12(A)(B)

Delmar, *Firefighter's Handbook*, 3rd Edition, 1st Printing, page 657.

IFSTA, *Essentials of Fire Fighting and Fire Department Operations*, 5th Edition, 1st Printing, page 560.

Jones and Bartlett, NFPA, *Fundamentals of Fire Fighter Skills*, 2nd Edition, 1st Printing, page 420.

Answer: B

101. Reference: NFPA 1001, 5.3.12 and 5.3.12(A)(B)

Delmar, *Firefighter's Handbook*, 3rd Edition, 1st Printing, pages 657–659.

IFSTA, *Essentials of Fire Fighting and Fire Department Operations*, 5th Edition, 1st Printing, pages 562–563.

Jones and Bartlett, NFPA, *Fundamentals of Fire Fighter Skills*, 2nd Edition, 1st Printing, page 427.

Answer: C

102. Reference: NFPA 1001, 5.3.13 and 5.3.13(A)(B)

Delmar, *Firefighter's Handbook*, 3rd Edition, 1st Printing, page 736.

IFSTA, *Essentials of Fire Fighting and Fire Department Operations*, 5th Edition, 1st Printing, page 881.

Jones and Bartlett, NFPA, *Fundamentals of Fire Fighter Skills*, 2nd Edition, 1st Printing, page 573.

Answer: D

103. Reference: NFPA 1001, 5.3.13 and 5.3.13(A)(B)

Delmar, *Firefighter's Handbook*, 3rd Edition, 1st Printing, page 743.

IFSTA, *Essentials of Fire Fighting and Fire Department Operations*, 5th Edition, 1st Printing, page 885.

Jones and Bartlett, NFPA, *Fundamentals of Fire Fighter Skills*, 2nd Edition, 1st Printing, page 578.

Answer: B

104. Reference: NFPA 1001, 5.3.13 and 5.3.13(A)(B)

Delmar, *Firefighter's Handbook*, 3rd Edition, 1st Printing, page 743.

IFSTA, *Essentials of Fire Fighting and Fire Department Operations*, 5th Edition, 1st Printing, page 884.

Jones and Bartlett, NFPA, *Fundamentals of Fire Fighter Skills*, 2nd Edition, 1st Printing, page 576.

Answer: A

105. Reference: NFPA 1001, 5.3.13 and 5.3.13(A)(B)

Delmar, *Firefighter's Handbook*, 3rd Edition, 1st Printing, pages 743–744.

IFSTA, *Essentials of Fire Fighting and Fire Department Operations*, 5th Edition, 1st Printing, page 869.

Jones and Bartlett, NFPA, *Fundamentals of Fire Fighter Skills*, 2nd Edition, 1st Printing, page 570–571.

Answer: C

106. Reference: NFPA 1001, 5.3.13, 5.3.13(A)(B), 5.3.14 and 5.3.14(A)(B)
Delmar, *Firefighter's Handbook*, 3rd Edition, 1st Printing, page 747.
IFSTA, *Essentials of Fire Fighting and Fire Department Operations*, 5th Edition, 1st Printing, page 916.
Jones and Bartlett, NFPA, *Fundamentals of Fire Fighter Skills*, 2nd Edition, 1st Printing, pages 575 and 973.
Answer: A

107. Reference: NFPA 1001, 5.3.14 and 5.3.14(A)(B)
Delmar, *Firefighter's Handbook*, 3rd Edition, 1st Printing, page 738.
IFSTA, *Essentials of Fire Fighting and Fire Department Operations*, 5th Edition, 1st Printing, pages 879 and 897.
Jones and Bartlett, NFPA, *Fundamentals of Fire Fighter Skills*, 2nd Edition, 1st Printing, page 568.
Answer: B

108. Reference: NFPA 1001, 5.3.14 and 5.3.14(A)
Delmar, *Firefighter's Handbook*, 3rd Edition, 1st Printing, page 727.
IFSTA, *Essentials of Fire Fighting and Fire Department Operations*, 5th Edition, 1st Printing, page 868.
Jones and Bartlett, NFPA, *Fundamentals of Fire Fighter Skills*, 2nd Edition, 1st Printing, page 559.
Answer: C

109. Reference: NFPA 1001, 5.3.14 and 5.3.14(A)(B)
Delmar, *Firefighter's Handbook*, 3rd Edition, 1st Printing, pages 740–741.
IFSTA, *Essentials of Fire Fighting and Fire Department Operations*, 5th Edition, 1st Printing, page 900, Skill sheet 17-1–10.
Jones and Bartlett, NFPA, *Fundamentals of Fire Fighter Skills*, 2nd Edition, 1st Printing, page 566.
Answer: C

110. Reference: NFPA 1001, 5.3.14 and 5.3.14(A)(B)
Delmar, *Firefighter's Handbook*, 3rd Edition, 1st Printing, page 743.
IFSTA, *Essentials of Fire Fighting and Fire Department Operations*, 5th Edition, 1st Printing, pages 876.
Jones and Bartlett, NFPA, *Fundamentals of Fire Fighter Skills*, 2nd Edition, 1st Printing, page 564.
Answer: D

111. Reference: NFPA 1001, 5.3.14 and 5.3.14(A)
Delmar, *Firefighter's Handbook*, 3rd Edition, 1st Printing, pages 366–367, 369.
IFSTA, *Essentials of Fire Fighting and Fire Department Operations*, 5th Edition, 1st Printing, pages 850–851.
Jones and Bartlett, NFPA, *Fundamentals of Fire Fighter Skills*, 2nd Edition, 1st Printing, page 959.
Answer: B

112. Reference: NFPA 1001, 5.3.14 and 5.3.14(A)(B)

Delmar, *Firefighter's Handbook*, 3rd Edition, 1st Printing, pages 370, 741–742.

IFSTA, *Essentials of Fire Fighting and Fire Department Operations*, 5th Edition, 1st Printing, pages 847 and 856.

Jones and Bartlett, NFPA, *Fundamentals of Fire Fighter Skills*, 2nd Edition, 1st Printing, page 561.

Answer: C

113. Reference: NFPA 1001, 5.3.14 and 5.3.14(A)(B)

Delmar, *Firefighter's Handbook*, 3rd Edition, 1st Printing, page 729.

IFSTA, *Essentials of Fire Fighting and Fire Department Operations*, 5th Edition, 1st Printing, page 849.

Jones and Bartlett, NFPA, *Fundamentals of Fire Fighter Skills*, 2nd Edition, 1st Printing, page 565.

Answer: B

114. Reference: NFPA 1001, 5.3.14 and 5.3.14(A)

Delmar, *Firefighter's Handbook*, 3rd Edition, 1st Printing, page 353.

IFSTA, *Essentials of Fire Fighting and Fire Department Operations*, 5th Edition, 1st Printing, page 829.

Jones and Bartlett, NFPA, *Fundamentals of Fire Fighter Skills*, 2nd Edition, 1st Printing, page 943.

Answer: B

115. Reference: NFPA 1001, 5.3.15 and 5.3.15(A)(B)

Delmar, *Firefighter's Handbook*, 3rd Edition, 1st Printing, page 239.

IFSTA, *Essentials of Fire Fighting and Fire Department Operations*, 5th Edition, 1st Printing, page 695 Skill sheet 13-1–12.

Jones and Bartlett, NFPA, *Fundamentals of Fire Fighter Skills*, 2nd Edition, 1st Printing, page 454.

Answer: B

116. Reference: NFPA 1001, 5.3.15 and 5.3.15(A)(B)

Delmar, *Firefighter's Handbook*, 3rd Edition, 1st Printing, page 311.

IFSTA, *Essentials of Fire Fighting and Fire Department Operations*, 5th Edition, 1st Printing, pages 664–665.

Jones and Bartlett, NFPA, *Fundamentals of Fire Fighter Skills*, 2nd Edition, 1st Printing, page 487.

Answer: B

117. Reference: NFPA 1001, 5.3.15 and 5.3.15(A)(B)

Delmar, *Firefighter's Handbook*, 3rd Edition, 1st Printing, page 312.

IFSTA, *Essentials of Fire Fighting and Fire Department Operations*, 5th Edition, 1st Printing, pages 666–667.

Jones and Bartlett, NFPA, *Fundamentals of Fire Fighter Skills*, 2nd Edition, 1st Printing, page 488.

Answer: B

118. Reference: NFPA 1001, 5.3.15 and 5.3.15(A)
Delmar, *Firefighter's Handbook*, 3rd Edition, 1st Printing, page 256.
IFSTA, *Essentials of Fire Fighting and Fire Department Operations*, 5th Edition, 1st Printing, page 633.
Jones and Bartlett, NFPA, *Fundamentals of Fire Fighter Skills*, 2nd Edition, 1st Printing, page 474.
Answer: A

119. Reference: NFPA 1001, 5.3.15 and 5.3.15(A)
Delmar, *Firefighter's Handbook*, 3rd Edition, 1st Printing, page 311.
IFSTA, *Essentials of Fire Fighting and Fire Department Operations*, 5th Edition, 1st Printing, page 664.
Jones and Bartlett, NFPA, *Fundamentals of Fire Fighter Skills*, 2nd Edition, 1st Printing, page 487.
Answer: D

120. Reference: NFPA 1001, 5.3.15 and 5.3.15(A)
Delmar, *Firefighter's Handbook*, 3rd Edition, 1st Printing, page 257.
IFSTA, *Essentials of Fire Fighting and Fire Department Operations*, 5th Edition, 1st Printing, page 655.
Jones and Bartlett, NFPA, *Fundamentals of Fire Fighter Skills*, 2nd Edition, 1st Printing, page 492.
Answer: C

121. Reference: NFPA 1001, 5.3.15 and 5.3.15(A)(B)
Delmar, *Firefighter's Handbook*, 3rd Edition, 1st Printing, pages 242–243.
IFSTA, *Essentials of Fire Fighting and Fire Department Operations*, 5th Edition, 1st Printing, page 615.
Jones and Bartlett, NFPA, *Fundamentals of Fire Fighter Skills*, 2nd Edition, 1st Printing, page 446.
Answer: A

122. Reference: NFPA 1001, 5.3.15 and 5.3.15(A)
Delmar, *Firefighter's Handbook*, 3rd Edition, 1st Printing, page 237.
IFSTA, *Essentials of Fire Fighting and Fire Department Operations*, 5th Edition, 1st Printing, page 597.
Jones and Bartlett, NFPA, *Fundamentals of Fire Fighter Skills*, 2nd Edition, 1st Printing, page 450.
Answer: B

123. Reference: NFPA 1001, 5.3.15 and 5.3.15(A)
Delmar, *Firefighter's Handbook*, 3rd Edition, 1st Printing, page 238.
IFSTA, *Essentials of Fire Fighting and Fire Department Operations*, 5th Edition, 1st Printing, pages 599–600.
Jones and Bartlett, NFPA, *Fundamentals of Fire Fighter Skills*, 2nd Edition, 1st Printing, page 450.
Answer: C

124. Reference: NFPA 1001, 5.3.15 and 5.3.15(A)

Delmar, *Firefighter's Handbook*, 3rd Edition, 1st Printing, pages 238–239.

IFSTA, *Essentials of Fire Fighting and Fire Department Operations*, 5th Edition, 1st Printing, pages 605–607.

Jones and Bartlett, NFPA, *Fundamentals of Fire Fighter Skills*, 2nd Edition, 1st Printing, pages 452–453.

Answer: D

125. Reference: NFPA 1001, 5.3.16 and 5.3.16(A)(B)

Delmar, *Firefighter's Handbook*, 3rd Edition, 1st Printing, pages 218–219.

IFSTA, *Essentials of Fire Fighting and Fire Department Operations*, 5th Edition, 1st Printing, page 249.

Jones and Bartlett, NFPA, *Fundamentals of Fire Fighter Skills*, 2nd Edition, 1st Printing, page 184.

Answer: C

126. Reference: NFPA 1001, 5.3.16 and 5.3.16(A)(B)

Delmar, *Firefighter's Handbook*, 3rd Edition, 1st Printing, page 215.

IFSTA, *Essentials of Fire Fighting and Fire Department Operations*, 5th Edition, 1st Printing, pages 110 and 249.

Jones and Bartlett, NFPA, *Fundamentals of Fire Fighter Skills*, 2nd Edition, 1st Printing, page 182.

Answer: A

127. Reference: NFPA 1001, 5.3.16 and 5.3.16(A)(B)

Delmar, *Firefighter's Handbook*, 3rd Edition, 1st Printing, page 216.

IFSTA, *Essentials of Fire Fighting and Fire Department Operations*, 5th Edition, 1st Printing, page 249.

Jones and Bartlett, NFPA, *Fundamentals of Fire Fighter Skills*, 2nd Edition, 1st Printing, pages 188–189.

Answer: C

128. Reference: NFPA 1001, 5.3.16 and 5.3.16(A)(B)

Delmar, *Firefighter's Handbook*, 3rd Edition, 1st Printing, page 219.

IFSTA, *Essentials of Fire Fighting and Fire Department Operations*, 5th Edition, 1st Printing, page 249.

Jones and Bartlett, NFPA, *Fundamentals of Fire Fighter Skills*, 2nd Edition, 1st Printing, page 184.

Answer: C

129. Reference: NFPA 1001, 5.3.16 and 5.3.16(A)(B)

Delmar, *Firefighter's Handbook*, 3rd Edition, 1st Printing, pages 217–218.

IFSTA, *Essentials of Fire Fighting and Fire Department Operations*, 5th Edition, 1st Printing, page 237.

Jones and Bartlett, NFPA, *Fundamentals of Fire Fighter Skills*, 2nd Edition, 1st Printing, page 188.

Answer: A

130. Reference: NFPA 1001, 5.3.16 and 5.3.16(A)(B)
Delmar, *Firefighter's Handbook*, 3rd Edition, 1st Printing, page 216.
IFSTA, *Essentials of Fire Fighting and Fire Department Operations*, 5th Edition, 1st Printing, page 237.
Jones and Bartlett, NFPA, *Fundamentals of Fire Fighter Skills*, 2nd Edition, 1st Printing, page 191.
Answer: D

131. Reference: NFPA 1001, 5.3.16 and 5.3.16(A)(B)
Delmar, *Firefighter's Handbook*, 3rd Edition, 1st Printing, page 218.
IFSTA, *Essentials of Fire Fighting and Fire Department Operations*, 5th Edition, 1st Printing, page 254.
Jones and Bartlett, NFPA, *Fundamentals of Fire Fighter Skills*, 2nd Edition, 1st Printing, page 190.
Answer: B

132. Reference: NFPA 1001, 5.3.16 and 5.3.16(A)(B)
Delmar, *Firefighter's Handbook*, 3rd Edition, 1st Printing, page 216.
IFSTA, *Essentials of Fire Fighting and Fire Department Operations*, 5th Edition, 1st Printing, page 243.
Jones and Bartlett, NFPA, *Fundamentals of Fire Fighter Skills*, 2nd Edition, 1st Printing, page 188.
Answer: A

133. Reference: NFPA 1001, 5.3.16 and 5.3.16(A)(B)
Delmar, *Firefighter's Handbook*, 3rd Edition, 1st Printing, pages 218–219.
IFSTA, *Essentials of Fire Fighting and Fire Department Operations*, 5th Edition, 1st Printing, page 236.
Jones and Bartlett, NFPA, *Fundamentals of Fire Fighter Skills*, 2nd Edition, 1st Printing, pages 197 and 198.
Answer: B

134. Reference: NFPA 1001, 5.3.16 and 5.3.16(A)(B)
Delmar, *Firefighter's Handbook*, 3rd Edition, 1st Printing, page 218.
IFSTA, *Essentials of Fire Fighting and Fire Department Operations*, 5th Edition, 1st Printing, page 781.
Jones and Bartlett, NFPA, *Fundamentals of Fire Fighter Skills*, 2nd Edition, 1st Printing, page 190.
Answer: B

135. Reference: NFPA 1001, 5.3.16 and 5.3.16(A)(B)
Delmar, *Firefighter's Handbook*, 3rd Edition, 1st Printing, page 218.
IFSTA, *Essentials of Fire Fighting and Fire Department Operations*, 5th Edition, 1st Printing, page 237.
Jones and Bartlett, NFPA, *Fundamentals of Fire Fighter Skills*, 2nd Edition, 1st Printing, page 193.
Answer: A

136. Reference: NFPA 1001, 5.3.16, and 5.3.16(A)(B)

Delmar, *Firefighter's Handbook*, 3rd Edition, 1st Printing, page 216.

IFSTA, *Essentials of Fire Fighting and Fire Department Operations*, 5th Edition, 1st Printing, page 237.

Jones and Bartlett, NFPA, *Fundamentals of Fire Fighter Skills*, 2nd Edition, 1st Printing, page 183.

Answer: C

137. Reference: NFPA 1001, 5.3.18 and 5.3.18(A)(B)

Delmar, *Firefighter's Handbook*, 3rd Edition, 1st Printing, page 574.

IFSTA, *Essentials of Fire Fighting and Fire Department Operations*, 5th Edition, 1st Printing, page 369.

Jones and Bartlett, NFPA, *Fundamentals of Fire Fighter Skills*, 2nd Edition, 1st Printing, page 637.

Answer: B

138. Reference: NFPA 1001, 5.3.18 and 5.3.18(A)(B)

Delmar, *Firefighter's Handbook*, 3rd Edition, 1st Printing, page 573.

IFSTA, *Essentials of Fire Fighting and Fire Department Operations*, 5th Edition, 1st Printing, page 369.

Jones and Bartlett, NFPA, *Fundamentals of Fire Fighter Skills*, 2nd Edition, 1st Printing, page 33.

Answer: A

139. Reference: NFPA 1001, 5.3.18 and 5.3.18(A)(B)

Delmar, *Firefighter's Handbook*, 3rd Edition, 1st Printing, page 574.

IFSTA, *Essentials of Fire Fighting and Fire Department Operations*, 5th Edition, 1st Printing, pages 783 and 811.

Jones and Bartlett, NFPA, *Fundamentals of Fire Fighter Skills*, 2nd Edition, 1st Printing, page 636.

Answer: C

140. Reference: NFPA 1001, 5.5.1 and 5.5.1(A)(B)

Delmar, *Firefighter's Handbook*, 3rd Edition, 1st Printing, page 422.

IFSTA, *Essentials of Fire Fighting and Fire Department Operations*, 5th Edition, 1st Printing, page 480.

Jones and Bartlett, NFPA, *Fundamentals of Fire Fighter Skills*, 2nd Edition, 1st Printing, page 338.

Answer: D

141. Reference: NFPA 1001, 5.5.1, and 5.5.1(A)

Delmar, *Firefighter's Handbook*, 3rd Edition, 1st Printing, pages 498–499.

IFSTA, *Essentials of Fire Fighting and Fire Department Operations*, 5th Edition, 1st Printing, page 271.

Jones and Bartlett, NFPA, *Fundamentals of Fire Fighter Skills*, 2nd Edition, 1st Printing, page 246.

Answer: C

142. Reference: NFPA 1001, 5.5.1 and 5.5.1(A)(B)

Delmar, *Firefighter's Handbook*, 3rd Edition, 1st Printing, page 499.

IFSTA, *Essentials of Fire Fighting and Fire Department Operations*, 5th Edition, 1st Printing, page 272.

Jones and Bartlett, NFPA, *Fundamentals of Fire Fighter Skills*, 2nd Edition, 1st Printing, page 246.

Answer: C

143. Reference: NFPA 1001, 5.5.1, and 5.5.1(A)(B)

Delmar, *Firefighter's Handbook*, 3rd Edition, 1st Printing, page 466.

IFSTA, *Essentials of Fire Fighting and Fire Department Operations*, 5th Edition, 1st Printing, page 269.

Jones and Bartlett, NFPA, *Fundamentals of Fire Fighter Skills*, 2nd Edition, 1st Printing, page 247.

Answer: C

144. Reference: NFPA 1001, 5.5.1 and 5.5.1(A)(B)

Delmar, *Firefighter's Handbook*, 3rd Edition, 1st Printing, pages 199, 201,202.

IFSTA, *Essentials of Fire Fighting and Fire Department Operations*, 5th Edition, 1st Printing, page 205.

Jones and Bartlett, NFPA, *Fundamentals of Fire Fighter Skills*, 2nd Edition, 1st Printing, page 64.

Answer: A

145. Reference: NFPA 1001, 5.5.1 and 5.5.1(A)(B)

Delmar, *Firefighter's Handbook*, 3rd Edition, 1st Printing, page 424.

IFSTA, *Essentials of Fire Fighting and Fire Department Operations*, 5th Edition, 1st Printing, page 481.

Jones and Bartlett, NFPA, *Fundamentals of Fire Fighter Skills*, 2nd Edition, 1st Printing, page 333.

Answer: D

146. Reference: NFPA 1001, 5.5.1 and 5.5.1(A)(B)

Delmar, *Firefighter's Handbook*, 3rd Edition, 1st Printing, page 598.

IFSTA, *Essentials of Fire Fighting and Fire Department Operations*, 5th Edition, 1st Printing, page 411.

Jones and Bartlett, NFPA, *Fundamentals of Fire Fighter Skills*, 2nd Edition, 1st Printing, page 231.

Answer: C

147. Reference: NFPA 1001, 5.5.2 and 5.5.2(A)(B)

Delmar, *Firefighter's Handbook*, 3rd Edition, 1st Printing, page 258.

IFSTA, *Essentials of Fire Fighting and Fire Department Operations*, 5th Edition, 1st Printing, page 637.

Jones and Bartlett, NFPA, *Fundamentals of Fire Fighter Skills*, 2nd Edition, 1st Printing, pages 476–477.

Answer: C

148. Reference: NFPA 1001, 5.5.2, 5.5.2(A)(B), 5.3.10 and 5.3.10(A)(B)
Delmar, *Firefighter's Handbook*, 3rd Edition, 1st Printing, page 266.
IFSTA, *Essentials of Fire Fighting and Fire Department Operations*, 5th Edition, 1st Printing, page 653.
Jones and Bartlett, NFPA, *Fundamentals of Fire Fighter Skills*, 2nd Edition, 1st Printing, page 482.
Answer: B

149. Reference: NFPA 1001, 5.5.2, 5.5.2(A)(B), 5.3.15 and 5.3.15(A)(B)
Delmar, *Firefighter's Handbook*, 3rd Edition, 1st Printing, page 312.
IFSTA, *Essentials of Fire Fighting and Fire Department Operations*, 5th Edition, 1st Printing, page 666.
Jones and Bartlett, NFPA, *Fundamentals of Fire Fighter Skills*, 2nd Edition, 1st Printing, page 488.
Answer: C

150. Reference: NFPA 1001, 5.5.2 and 5.5.2(A)(B)
Delmar, *Firefighter's Handbook*, 3rd Edition, 1st Printing, pages 257–258.
IFSTA, *Essentials of Fire Fighting and Fire Department Operations*, 5th Edition, 1st Printing, page 634.
Jones and Bartlett, NFPA, *Fundamentals of Fire Fighter Skills*, 2nd Edition, 1st Printing, pages 466–476.
Answer: B

Don't forget to enter the information on your Personal Progress Plotter and answer the Yes and No question at the end of the Examination. This step is extremely important for the successful completion of the Systematic Approach to Examination Preparation!

Examination I-3 Answer Key

Directions

Follow these steps carefully for completing the feedback part of SAEP:

1. After calculating your score, look up the answers for the examination items you missed as well as those on which you guessed, even if you guessed correctly. If you are guessing, it means the answer is not perfectly clear. In this process, we are committed to making you as knowledgeable as possible.

2. Enter the number of missed and guessed examination items in the blanks on your Personal Progress Plotter.

3. Highlight the answer in the reference materials. Read the paragraph preceding and the paragraph following the one in which the correct answer is located. Enter the paragraph number and page number next to the guessed or missed examination item on your examination. Count any part of a paragraph at the beginning of the page as one paragraph until you reach the paragraph containing your highlighted answer. This step will help you locate and review your missed and guessed examination items later in the process. This step is essential to learning the material in context and by association. These learning techniques (context/association) are the very backbone of the SAEP approach.

4. Congratulations! You have completed the examination and feedback steps of SAEP when you have highlighted your guessed and missed examination items for this examination.

Proceed to Phases III and IV. Study the materials carefully in these important phases—they will help you polish your examination-taking skills. Approximately two to three days before you take your next examination, carefully read all of the highlighted information in the reference materials using the same techniques you applied during the feedback step. This will reinforce your learning and provide you with an added level of confidence going into the examination.

Someone once said to professional golfer Tom Watson after he won several tournament championships, "You are really lucky to have won those championships. You are really on a streak." Watson was reported to have replied, "Yes, there is some luck involved, but what I have really noticed is that the more I practice, the luckier I get." What Watson was saying is that good luck usually results from good preparation. This line of thinking certainly applies to learning the rules and hints of examination taking.

──────────── **Rule 7** ────────────

Good luck = good preparation.

1. Reference: NFPA 1001, 5.1.1, 4.2
   Delmar, *Firefighter's Handbook*, 3rd Edition, 1st Printing, page 120.
   IFSTA, *Essentials of Fire Fighting and Fire Department Operations*, 5th Edition, 1st Printing, page 65.
   Jones and Bartlett, NFPA, *Fundamentals of Fire Fighter Skills*, 2nd Edition, 1st Printing, page 25.
   Answer: C

2. Reference: NFPA 1001, 5.1.1
Delmar, *Firefighter's Handbook*, 3rd Edition, 1st Printing, page 37.
IFSTA, *Essentials of Fire Fighting and Fire Department Operations*, 5th Edition, 1st Printing, page 33.
Jones and Bartlett, NFPA, *Fundamentals of Fire Fighter Skills*, 2nd Edition, 1st Printing, page 6.
Answer: B

3. Reference: NFPA 1001, 5.1.1, 5.5.1, and 5.5.1(A)
Delmar, *Firefighter's Handbook*, 3rd Edition, 1st Printing, page 476.
IFSTA, *Essentials of Fire Fighting and Fire Department Operations*, 5th Edition, 1st Printing, page 267.
Jones and Bartlett, NFPA, *Fundamentals of Fire Fighter Skills*, 2nd Edition, 1st Printing, page 242.
Answer: B

4. Reference: NFPA 1001, 5.1.1, 5.5.1, and 5.5.1(A)
Delmar, *Firefighter's Handbook*, 3rd Edition, 1st Printing, page 499.
IFSTA, *Essentials of Fire Fighting and Fire Department Operations*, 5th Edition, 1st Printing, page 272.
Jones and Bartlett, NFPA, *Fundamentals of Fire Fighter Skills*, 2nd Edition, 1st Printing, page 246.
Answer: A

5. Reference: NFPA 1001, 5.1.1 and 5.1.2
Delmar, *Firefighter's Handbook*, 3rd Edition, 1st Printing, page 481.
IFSTA, *Essentials of Fire Fighting and Fire Department Operations*, 5th Edition, 1st Printing, page 275.
Jones and Bartlett, NFPA, *Fundamentals of Fire Fighter Skills*, 2nd Edition, 1st Printing, page 248.
Answer: C

6. Reference: NFPA 1001, 5.1.1
Delmar, *Firefighter's Handbook*, 3rd Edition, 1st Printing, page 42.
IFSTA, *Essentials of Fire Fighting and Fire Department Operations*, 5th Edition, 1st Printing, page 792.
Jones and Bartlett, NFPA, *Fundamentals of Fire Fighter Skills*, 2nd Edition, 1st Printing, page 589.
Answer: C

7. Reference: NFPA 1001, 5.1.1
Delmar, *Firefighter's Handbook*, 3rd Edition, 1st Printing, page 840.
IFSTA, *Essentials of Fire Fighting and Fire Department Operations*, 5th Edition, 1st Printing, page 62.
Jones and Bartlett, NFPA, *Fundamentals of Fire Fighter Skills*, 2nd Edition, 1st Printing, page 34.
Answer: A

8. Reference: NFPA 1001, 5.1.1, and 5.1.2
Delmar, *Firefighter's Handbook*, 3rd Edition, 1st Printing, page 481.
IFSTA, *Essentials of Fire Fighting and Fire Department Operations*, 5th Edition,
1st Printing, page 276.
Jones and Bartlett, NFPA, *Fundamentals of Fire Fighter Skills*, 2nd Edition,
1st Printing, page 251.
Answer: A

9. Reference: NFPA 1001, 5.1.1, and 5.1.2
Delmar, *Firefighter's Handbook*, 3rd Edition, 1st Printing, page 484.
IFSTA, *Essentials of Fire Fighting and Fire Department Operations*, 5th Edition,
1st Printing, page 293.
Jones and Bartlett, NFPA, *Fundamentals of Fire Fighter Skills*, 2nd Edition,
1st Printing, page 262.
Answer: B

10. Reference: NFPA 1001, 5.1.1 and 5.1.2
Delmar, *Firefighter's Handbook*, 3rd Edition, 1st Printing, page 478.
IFSTA, *Essentials of Fire Fighting and Fire Department Operations*, 5th Edition,
1st Printing, page 274.
Jones and Bartlett, NFPA, *Fundamentals of Fire Fighter Skills*, 2nd Edition,
1st Printing, page 248.
Answer: D

11. Reference: NFPA 1001, 5.1.1, and 5.1.2
Delmar, *Firefighter's Handbook*, 3rd Edition, 1st Printing, page 479.
IFSTA, *Essentials of Fire Fighting and Fire Department Operations*, 5th Edition,
1st Printing, page 275.
Jones and Bartlett, NFPA, *Fundamentals of Fire Fighter Skills*, 2nd Edition,
1st Printing, page 249.
Answer: C

12. Reference: NFPA 1001, 5.1.1 and 5.1.2
Delmar, *Firefighter's Handbook*, 3rd Edition, 1st Printing, page 473.
IFSTA, *Essentials of Fire Fighting and Fire Department Operations*, 5th Edition,
1st Printing, pages 265 and 266.
Jones and Bartlett, NFPA, *Fundamentals of Fire Fighter Skills*, 2nd Edition,
1st Printing, page 240.
Answer: B

13. Reference: NFPA 1001, 5.1.1, and 5.1.2
Delmar, *Firefighter's Handbook*, 3rd Edition, 1st Printing, pages 483–484.
IFSTA, *Essentials of Fire Fighting and Fire Department Operations*, 5th Edition,
1st Printing, page 293.
Jones and Bartlett, NFPA, *Fundamentals of Fire Fighter Skills*, 2nd Edition,
1st Printing, page 262.
Answer: B

14. Reference: NFPA 1001, 5.1.1

Delmar, *Firefighter's Handbook*, 3rd Edition, 1st Printing, page 695.

IFSTA, *Essentials of Fire Fighting and Fire Department Operations*, 5th Edition, 1st Printing, page 70.

Jones and Bartlett, NFPA, *Fundamentals of Fire Fighter Skills*, 2nd Edition, 1st Printing, pages 286–287 and 617.

Answer: A

15. Reference: NFPA 1001, 5.1.1

Delmar, *Firefighter's Handbook*, 3rd Edition, 1st Printing, pages 689, 908, and 1117.

IFSTA, *Essentials of Fire Fighting and Fire Department Operations*, 5th Edition, 1st Printing, page 94.

Jones and Bartlett, NFPA, *Fundamentals of Fire Fighter Skills*, 2nd Edition, 1st Printing, page 791.

Answer: A

16. Reference: NFPA 1001, 5.2.1 and 5.2.1(A)(B)

Delmar, *Firefighter's Handbook*, 3rd Edition, 1st Printing, page 72.

IFSTA, *Essentials of Fire Fighting and Fire Department Operations*, 5th Edition, 1st Printing, page 943.

Jones and Bartlett, NFPA, *Fundamentals of Fire Fighter Skills*, 2nd Edition, 1st Printing, page 91.

Answer: C

17. Reference: NFPA 1001, 5.2.1 and 5.2.1(A)(B)

Delmar, *Firefighter's Handbook*, 3rd Edition, 1st Printing, pages 58 and 1105.

IFSTA, *Essentials of Fire Fighting and Fire Department Operations*, 5th Edition, 1st Printing, page 934.

Jones and Bartlett, NFPA, *Fundamentals of Fire Fighter Skills*, 2nd Edition, 1st Printing, page 81.

Answer: A

18. Reference: NFPA 1001, 5.2.1 and 5.2.1(A)(B)

Delmar, *Firefighter's Handbook*, 3rd Edition, 1st Printing, page 60.

IFSTA, *Essentials of Fire Fighting and Fire Department Operations*, 5th Edition, 1st Printing, page 936.

Jones and Bartlett, NFPA, *Fundamentals of Fire Fighter Skills*, 2nd Edition, 1st Printing, page 83.

Answer: D

19. Reference: NFPA 1001, 5.2.1 and 5.2.1(A)(B)

Delmar, *Firefighter's Handbook*, 3rd Edition, 1st Printing, page 61.

IFSTA, *Essentials of Fire Fighting and Fire Department Operations*, 5th Edition, 1st Printing, 938.

Jones and Bartlett, NFPA, *Fundamentals of Fire Fighter Skills*, 2nd Edition, 1st Printing, page 85.

Answer: C

20. Reference: NFPA 1001, 5.2.1, 5.2.1(A)(B), 5.2.2, 5.2.2(A)(B), 5.2.3, and 5.2.3(A)(B)
Delmar, *Firefighter's Handbook*, 3rd Edition, 1st Printing, page 55.
IFSTA, *Essentials of Fire Fighting and Fire Department Operations*, 5th Edition, 1st Printing, page 927.
Jones and Bartlett, NFPA, *Fundamentals of Fire Fighter Skills*, 2nd Edition, 1st Printing, page 82.
Answer: A

21. Reference: NFPA 1001, 5.3.1 and 5.3.1(A)(B)
Delmar, *Firefighter's Handbook*, 3rd Edition, 1st Printing, page 202.
IFSTA, *Essentials of Fire Fighting and Fire Department Operations*, 5th Edition, 1st Printing, page 229.
Jones and Bartlett, NFPA, *Fundamentals of Fire Fighter Skills*, 2nd Edition, 1st Printing, page 93.
Answer: B

22. Reference: NFPA 1001, 5.3.1 and 5.3.1(A)(B)
Delmar, *Firefighter's Handbook*, 3rd Edition, 1st Printing, page 188.
IFSTA, *Essentials of Fire Fighting and Fire Department Operations*, 5th Edition, 1st Printing, page 203.
Jones and Bartlett, NFPA, *Fundamentals of Fire Fighter Skills*, 2nd Edition, 1st Printing, page 53.
Answer: B

23. Reference: NFPA 1001, 5.3.1, 5.3.1(A)(B), 5.3.5, and 5.3.5(A)(B)
Delmar, *Firefighter's Handbook*, 3rd Edition, 1st Printing, page 162.
IFSTA, *Essentials of Fire Fighting and Fire Department Operations*, 5th Edition, 1st Printing, page 181.
Jones and Bartlett, NFPA, *Fundamentals of Fire Fighter Skills*, 2nd Edition, 1st Printing, page 48.
Answer: D

24. Reference: NFPA 1001, 5.3.1, 5.3.1(A)(B), 5.3.5, 5.3.5(A)(B), 5.3.10, 5.3.10(A), 5.3.11, and 5.3.11(A)
Delmar, *Firefighter's Handbook*, 3rd Edition, 1st Printing, page 162.
IFSTA, *Essentials of Fire Fighting and Fire Department Operations*, 5th Edition, 1st Printing, page 181.
Jones and Bartlett, NFPA, *Fundamentals of Fire Fighter Skills*, 2nd Edition, 1st Printing, page 48.
Answer: C

25. Reference: NFPA 1001, 5.3.1, 5.3.1(A), 5.3.5, 5.3.5(A)(B), 5.3.9, and 5.3.9(A)(B)
Delmar, *Firefighter's Handbook*, 3rd Edition, 1st Printing, page 573.
IFSTA, *Essentials of Fire Fighting and Fire Department Operations*, 5th Edition, 1st Printing, page 186.
Jones and Bartlett, NFPA, *Fundamentals of Fire Fighter Skills*, 2nd Edition, 1st Printing, page 48.
Answer: D

26. Reference: NFPA 1001, 5.3.1, 5.3.1(A), 5.3.5, 5.3.5(A)(B), 5.3.9, 5.3.9(A)(B), 5.3.10, 5.3.10(A), 5.3.11 and 5.3.11(A)

Delmar, *Firefighter's Handbook*, 3rd Edition, 1st Printing, pages 163–164.

IFSTA, *Essentials of Fire Fighting and Fire Department Operations*, 5th Edition, 1st Printing, page 182.

Jones and Bartlett, NFPA, *Fundamentals of Fire Fighter Skills*, 2nd Edition, 1st Printing, page 48.

Answer: C

27. Reference: NFPA 1001, 5.3.1, 5.3.1(A)(B), 5.3.5, and 5.3.5(A)(B)

Delmar, *Firefighter's Handbook*, 3rd Edition, 1st Printing, pages 127 and 167.

IFSTA, *Essentials of Fire Fighting and Fire Department Operations*, 5th Edition, 1st Printing, pages 187–188.

Jones and Bartlett, NFPA, *Fundamentals of Fire Fighter Skills*, 2nd Edition, 1st Printing, pages 50–51.

Answer: D

28. Reference: NFPA 1001, 5.3.1 and 5.3.1(A)(B)

Delmar, *Firefighter's Handbook*, 3rd Edition, 1st Printing, page 206.

IFSTA, *Essentials of Fire Fighting and Fire Department Operations*, 5th Edition, 1st Printing, page 206.

Jones and Bartlett, NFPA, *Fundamentals of Fire Fighter Skills*, 2nd Edition, 1st Printing, page 69.

Answer: A

29. Reference: NFPA 1001, 5.3.1, 5.3.1(A), 5.3.5, 5.3.5(A)(B), 5.3.11, and 5.3.11(A)

Delmar, *Firefighter's Handbook*, 3rd Edition, 1st Printing, page 161.

IFSTA, *Essentials of Fire Fighting and Fire Department Operations*, 5th Edition, 1st Printing, page 182.

Jones and Bartlett, NFPA, *Fundamentals of Fire Fighter Skills*, 2nd Edition, 1st Printing, page 131.

Answer: A

30. Reference: NFPA 1001, 5.3.1 and 5.3.1(A)(B)

Delmar, *Firefighter's Handbook*, 3rd Edition, 1st Printing, page 162.

IFSTA, *Essentials of Fire Fighting and Fire Department Operations*, 5th Edition, 1st Printing, page 180.

Jones and Bartlett, NFPA, *Fundamentals of Fire Fighter Skills*, 2nd Edition, 1st Printing, page 48.

Answer: B

31. Reference: NFPA 1001, 5.3.1 and 5.3.1(A)(B)

Delmar, *Firefighter's Handbook*, 3rd Edition, 1st Printing, page 206.

IFSTA, *Essentials of Fire Fighting and Fire Department Operations*, 5th Edition, 1st Printing, page 206.

Jones and Bartlett, NFPA, *Fundamentals of Fire Fighter Skills*, 2nd Edition, 1st Printing, page 67.

Answer: B

32. Reference: NFPA 1001, 5.3.1 and 5.3.1(A)(B)
Delmar, *Firefighter's Handbook*, 3rd Edition, 1st Printing, page 177.
IFSTA, *Essentials of Fire Fighting and Fire Department Operations*, 5th Edition,
1st Printing, page 189.
Jones and Bartlett, NFPA, *Fundamentals of Fire Fighter Skills*, 2nd Edition,
1st Printing, page 52.
Answer: C

33. Reference: NFPA 1001, 5.3.1 and 5.3.1(A)(B)
Delmar, *Firefighter's Handbook*, 3rd Edition, 1st Printing, page 179.
IFSTA, *Essentials of Fire Fighting and Fire Department Operations*, 5th Edition,
1st Printing, page 200.
Jones and Bartlett, NFPA, *Fundamentals of Fire Fighter Skills*, 2nd Edition,
1st Printing, page 56.
Answer: A

34. Reference: NFPA 1001, 5.3.1 and 5.3.1(A)(B)
Delmar, *Firefighter's Handbook*, 3rd Edition, 1st Printing, pages 177 and 186.
IFSTA, *Essentials of Fire Fighting and Fire Department Operations*, 5th Edition,
1st Printing, page 203.
Jones and Bartlett, NFPA, *Fundamentals of Fire Fighter Skills*, 2nd Edition,
1st Printing, page 53.
Answer: B

35. Reference: NFPA 1001, 5.3.2, 5.3.2(A), 5.3.3, and 5.3.3(A)(B)
Delmar, *Firefighter's Handbook*, 3rd Edition, 1st Printing, page 142.
IFSTA, *Essentials of Fire Fighting and Fire Department Operations*, 5th Edition,
1st Printing, page 175.
Jones and Bartlett, NFPA, *Fundamentals of Fire Fighter Skills*, 2nd Edition,
1st Printing, page 38.
Answer: A

36. Reference: NFPA 1001, 5.3.2 and 5.3.2(A)(B)
Delmar, *Firefighter's Handbook*, 3rd Edition, 1st Printing, page 143.
IFSTA, *Essentials of Fire Fighting and Fire Department Operations*, 5th Edition,
1st Printing, page 176.
Jones and Bartlett, NFPA, *Fundamentals of Fire Fighter Skills*, 2nd Edition,
1st Printing, page 39.
Answer: D

37. Reference: NFPA 1001, 5.3.2 and 5.3.2(A)(B)
Delmar, *Firefighter's Handbook*, 3rd Edition, 1st Printing, page 138.
IFSTA, *Essentials of Fire Fighting and Fire Department Operations*, 5th Edition,
1st Printing, page 167.
Jones and Bartlett, NFPA, *Fundamentals of Fire Fighter Skills*, 2nd Edition,
1st Printing, page 35.
Answer: B

38. Reference: NFPA 1001, 5.3.3 and 5.3.3(A)

Delmar, *Firefighter's Handbook*, 3rd Edition, 1st Printing, page 398.

IFSTA, *Essentials of Fire Fighting and Fire Department Operations*. 5th Edition, page 138.

Jones and Bartlett, NFPA, *Fundamentals of Fire Fighter Skills*, 2nd Edition, 1st Printing, page 162.

Answer: D

39. Reference: NFPA 1001, 5.3.4 and 5.3.4(A)(B)

Delmar, *Firefighter's Handbook*, 3rd Edition, 1st Printing, page 600.

IFSTA, *Essentials of Fire Fighting and Fire Department Operations*, 5th Edition, 1st Printing, page 433.

Jones and Bartlett, NFPA, *Fundamentals of Fire Fighter Skills*, 2nd Edition, 1st Printing, page 305.

Answer: B

40. Reference: NFPA 1001, 5.3.4 and 5.3.4(A)

Delmar, *Firefighter's Handbook*, 3rd Edition, 1st Printing, page 601.

IFSTA, *Essentials of Fire Fighting and Fire Department Operations*, 5th Edition, 1st Printing, page 417.

Jones and Bartlett, NFPA, *Fundamentals of Fire Fighter Skills*, 2nd Edition, 1st Printing, page 301.

Answer: C

41. Reference: NFPA 1001, 5.3.4 and 5.3.4(A)

Delmar, *Firefighter's Handbook*, 3rd Edition, 1st Printing, page 598.

IFSTA, *Essentials of Fire Fighting and Fire Department Operations*, 5th Edition, 1st Printing, page 415.

Jones and Bartlett, NFPA, *Fundamentals of Fire Fighter Skills*, 2nd Edition, 1st Printing, page 298.

Answer: A

42. Reference: NFPA 1001, 5.3.4 and 5.3.4(A)(B)

Delmar, *Firefighter's Handbook*, 3rd Edition, 1st Printing, pages 589–590.

IFSTA, *Essentials of Fire Fighting and Fire Department Operations*, 5th Edition, 1st Printing, page 446.

Jones and Bartlett, NFPA, *Fundamentals of Fire Fighter Skills*, 2nd Edition, 1st Printing, pages 222 and 294.

Answer: C

43. Reference: NFPA 1001, 5.3.4 and 5.3.4(A)(B)

Delmar, *Firefighter's Handbook*, 3rd Edition, 1st Printing, page 590.

IFSTA, *Essentials of Fire Fighting and Fire Department Operations*, 5th Edition, page 404.

Jones and Bartlett, NFPA, *Fundamentals of Fire Fighter Skills*, 2nd Edition, 1st Printing, page 219.

Answer: D

44. Reference: NFPA 1001, 5.3.4 and 5.3.4(A)(B)

Delmar, *Firefighter's Handbook*, 3rd Edition, 1st Printing, page 591.

IFSTA, *Essentials of Fire Fighting and Fire Department Operations*, 5th Edition, 1st Printing, page 430.

Jones and Bartlett, NFPA, *Fundamentals of Fire Fighter Skills*, 2nd Edition, 1st Printing, page 219.

Answer: C

45. Reference: NFPA 1001, 5.3.4 and 5.3.4(A)

Delmar, *Firefighter's Handbook*, 3rd Edition, 1st Printing, pages 593–594.

IFSTA, *Essentials of Fire Fighting and Fire Department Operations*, 5th Edition, 1st Printing, page 399.

Jones and Bartlett, NFPA, *Fundamentals of Fire Fighter Skills*, 2nd Edition, 1st Printing, page 226.

Answer: D

46. Reference: NFPA 1001, 5.3.4 and 5.3.4(A)(B)

Delmar, *Firefighter's Handbook*, 3rd Edition, 1st Printing, pages 594–595.

IFSTA, *Essentials of Fire Fighting and Fire Department Operations*, 5th Edition, 1st Printing, pages 434–435.

Jones and Bartlett, NFPA, *Fundamentals of Fire Fighter Skills*, 2nd Edition, 1st Printing, page 296.

Answer: A

47. Reference: NFPA 1001, 5.3.4 and 5.3.4(A)(B)

Delmar, *Firefighter's Handbook*, 3rd Edition, 1st Printing, page 603.

IFSTA, *Essentials of Fire Fighting and Fire Department Operations*, 5th Edition, 1st Printing, page 423.

Jones and Bartlett, NFPA, *Fundamentals of Fire Fighter Skills*, 2nd Edition, 1st Printing, page 302.

Answer: A

48. Reference: NFPA 1001, 5.3.4 and 5.3.4(A)(B)

Delmar, *Firefighter's Handbook*, 3rd Edition, pages 589–590.

IFSTA, *Essentials of Fire Fighting and Fire Department Operations*, 5th Edition, 1st Printing, page 446.

Jones and Bartlett, NFPA, *Fundamentals of Fire Fighter Skills*, 2nd Edition, 1st Printing, page 294.

Answer: A

49. Reference: NFPA 1001, 5.3.4 and 5.3.4(A)(B)

Delmar, *Firefighter's Handbook*, 3rd Edition, 1st Printing, pages 600–601.

IFSTA, *Essentials of Fire Fighting and Fire Department Operations*, 5th Edition, 1st Printing, page 413.

Jones and Bartlett, NFPA, *Fundamentals of Fire Fighter Skills*, 2nd Edition, 1st Printing, page 298.

Answer: D

50. Reference: NFPA 1001, 5.3.4 and 5.3.4(A)(B)

Delmar, *Firefighter's Handbook*, 3rd Edition, 1st Printing, pages 604–605.

IFSTA, *Essentials of Fire Fighting and Fire Department Operations*, 5th Edition, 1st Printing, page 426.

Jones and Bartlett, NFPA, *Fundamentals of Fire Fighter Skills*, 2nd Edition, 1st Printing, page 310.

Answer: D

51. Reference: NFPA 1001, 5.3.5 and 5.3.5(A)

Delmar, *Firefighter's Handbook*, 3rd Edition, 1st Printing, pages 824–825.

IFSTA, *Essentials of Fire Fighting and Fire Department Operations*, 5th Edition, 1st Printing, page 74.

Jones and Bartlett, NFPA, *Fundamentals of Fire Fighter Skills*, 2nd Edition, 1st Printing, pages 32, 278, and 534.

Answer: A

52. Reference: NFPA 1001, 5.3.5 and 5.3.5(A)

Delmar, *Firefighter's Handbook*, 3rd Edition, page 836.

IFSTA, *Essentials of Fire Fighting and Fire Department Operations*, 5th Edition, 1st Printing, page 71.

Jones and Bartlett, NFPA, *Fundamentals of Fire Fighter Skills*, 2nd Edition, 1st Printing, page 535.

Answer: C

53. Reference: NFPA 1001, 5.3.5 and 5.3.5(A)(B)

Delmar, *Firefighter's Handbook*, 3rd Edition, 1st Printing, page 121.

IFSTA, *Essentials of Fire Fighting and Fire Department Operations*, 5th Edition, 1st Printing, pages 77–78.

Jones and Bartlett, NFPA, *Fundamentals of Fire Fighter Skills*, 2nd Edition, 1st Printing, page 378.

Answer: A

54. Reference: NFPA 1001, 5.3.5 and 5.3.5(A)(B)

Delmar, *Firefighter's Handbook*, 3rd Edition, 1st Printing, pages 523–524.

IFSTA, *Essentials of Fire Fighting and Fire Department Operations*, 5th Edition, 1st Printing, page 209.

Jones and Bartlett, NFPA, *Fundamentals of Fire Fighter Skills*, 2nd Edition, 1st Printing, pages 377–378.

Answer: A

55. References: NFPA 1001, 5.3.6 and 5.3.6(A)(B)

Delmar, *Firefighter's Handbook*, 3rd Edition, 1st Printing, page 417.

IFSTA, *Essentials of Fire Fighting and Fire Department Operations*, 5th Edition, 1st Printing, page 475.

Jones and Bartlett, NFPA, *Fundamentals of Fire Fighter Skills*, 2nd Edition, 1st Printing, page 329.

Answer: D

56. References: NFPA 1001, 5.3.6 and 5.3.6(A)

Delmar, *Firefighter's Handbook*, 3rd Edition, 1st Printing, page 420.

IFSTA, *Essentials of Fire Fighting and Fire Department Operations*, 5th Edition, 1st Printing, page 476.

Jones and Bartlett, NFPA, *Fundamentals of Fire Fighter Skills*, 2nd Edition, 1st Printing, page 329.

Answer: A

57. References: NFPA 1001, 5.3.6 and 5.3.6(A)(B)

Delmar, *Firefighter's Handbook*, 3rd Edition, 1st Printing, pages 434–435 and 455.

IFSTA, *Essentials of Fire Fighting and Fire Department Operations*, 5th Edition, 1st Printing, pages 504–505, and 531.

Jones and Bartlett, NFPA, *Fundamentals of Fire Fighter Skills*, 2nd Edition, 1st Printing, page 357.

Answer: D

58. References: NFPA 1001, 5.3.6 and 5.3.6(A)

Delmar, *Firefighter's Handbook*, 3rd Edition, 1st Printing, page 417.

IFSTA, *Essentials of Fire Fighting and Fire Department Operations*, 5th Edition, 1st Printing, pages 472 and 474.

Jones and Bartlett, NFPA, *Fundamentals of Fire Fighter Skills*, 2nd Edition, 1st Printing, page 328.

Answer: C

59. References: NFPA 1001, 5.3.6, 5.3.6(A)(B), 5.3.9, 5.3.9(A)(B), 5.3.11, and 5.3.11(A)(B)

Delmar, *Firefighter's Handbook*, 3rd Edition, 1st Printing, pages 442–443.

IFSTA, *Essentials of Fire Fighting and Fire Department Operations*, 5th Edition, 1st Printing, pages 483 and 494.

Jones and Bartlett, NFPA, *Fundamentals of Fire Fighter Skills*, 2nd Edition, 1st Printing, page 347.

Answer: B

60. References: NFPA 1001, 5.3.6 and 5.3.6(A)(B)

Delmar, *Firefighter's Handbook*, 3rd Edition, 1st Printing, page 427.

IFSTA, *Essentials of Fire Fighting and Fire Department Operations*, 5th Edition, 1st Printing, pages 483 and 494.

Jones and Bartlett, NFPA, *Fundamentals of Fire Fighter Skills*, 2nd Edition, 1st Printing, page 347.

Answer: B

61. Reference: NFPA 1001, 5.3.6 and 5.3.6(A)

Delmar, *Firefighter's Handbook*, 3rd Edition, 1st Printing, page 417.

IFSTA, *Essentials of Fire Fighting and Fire Department Operations*, 5th Edition, 1st Printing, pages 473 and 475.

Jones and Bartlett, NFPA, *Fundamentals of Fire Fighter Skills*, 2nd Edition, 1st Printing, page 328.

Answer: D

62. Reference: NFPA 1001, 5.3.6 and 5.3.6(A)

Delmar, *Firefighter's Handbook*, 3rd Edition, 1st Printing, pages 422 and 425.

IFSTA, *Essentials of Fire Fighting and Fire Department Operations*, 5th Edition, 1st Printing, page 479.

Jones and Bartlett, NFPA, *Fundamentals of Fire Fighter Skills*, 2nd Edition, 1st Printing, page 333.

Answer: D

63. Reference: NFPA 1001, 5.3.6 and 5.3.6(A)(B)

Delmar, *Firefighter's Handbook*, 3rd Edition, 1st Printing, pages 427–428.

IFSTA, *Essentials of Fire Fighting and Fire Department Operations*, 5th Edition, 1st Printing, page 491.

Jones and Bartlett, NFPA, *Fundamentals of Fire Fighter Skills*, 2nd Edition, 1st Printing, page 339.

Answer: C

64. Reference: NFPA 1001, 5.3.7 and 5.3.7(A)(B)

Delmar, *Firefighter's Handbook*, 3rd Edition, 1st Printing, page 714.

IFSTA, *Essentials of Fire Fighting and Fire Department Operations*, 5th Edition, 1st Printing, page 799.

Jones and Bartlett, NFPA, *Fundamentals of Fire Fighter Skills*, 2nd Edition, 1st Printing, pages 631–632.

Answer: A

65. Reference: NFPA 1001, 5.3.9 and 5.3.9(A)(B)

Delmar, *Firefighter's Handbook*, 3rd Edition, 1st Printing, page 524.

IFSTA, *Essentials of Fire Fighting and Fire Department Operations*, 5th Edition, 1st Printing, pages 309–310.

Jones and Bartlett, NFPA, *Fundamentals of Fire Fighter Skills*, 2nd Edition, 1st Printing, page 370.

Answer: A

66. Reference: NFPA 1001, 5.3.9 and 5.3.9(A)(B)

Delmar, *Firefighter's Handbook*, 3rd Edition, 1st Printing, pages 521–522.

IFSTA, *Essentials of Fire Fighting and Fire Department Operations*, 5th Edition, 1st Printing, page 316.

Jones and Bartlett, NFPA, *Fundamentals of Fire Fighter Skills*, 2nd Edition, 1st Printing, page 371.

Answer: B

67. Reference: NFPA 1001, 5.3.9 and 5.3.9(A)(B)

Delmar, *Firefighter's Handbook*, 3rd Edition, 1st Printing, page 533.

IFSTA, *Essentials of Fire Fighting and Fire Department Operations*, 5th Edition, 1st Printing, page 328.

Jones and Bartlett, NFPA, *Fundamentals of Fire Fighter Skills*, 2nd Edition, 1st Printing, page 382.

Answer: B

68. Reference: NFPA 1001, 5.3.10 and 5.3.10(A)(B)

Delmar, *Firefighter's Handbook*, 3rd Edition, 1st Printing, pages 377 and 705.

IFSTA, *Essentials of Fire Fighting and Fire Department Operations*, 5th Edition, 1st Printing, page 674.

Jones and Bartlett, NFPA, *Fundamentals of Fire Fighter Skills*, 2nd Edition, 1st Printing, page 514.

Answer: A

69. Reference: NFPA 1001, 5.3.10 and 5.3.10(A)

Delmar, *Firefighter's Handbook*, 3rd Edition, 1st Printing, page 326.

IFSTA, *Essentials of Fire Fighting and Fire Department Operations*, 5th Edition, 1st Printing, page 727.

Jones and Bartlett, NFPA, *Fundamentals of Fire Fighter Skills*, 2nd Edition, 1st Printing, page 619.

Answer: D

70. Reference: NFPA 1001, 5.3.10, 5.3.10(A)(B), 5.3.8 and 5.3.8(A)(B)

Delmar, *Firefighter's Handbook*, 3rd Edition, 1st Printing, page 325.

IFSTA, *Essentials of Fire Fighting and Fire Department Operations*, 5th Edition, 1st Printing, page 764.

Jones and Bartlett, NFPA, *Fundamentals of Fire Fighter Skills*, 2nd Edition, 1st Printing, page 617.

Answer: C

71. Reference: NFPA 1001, 5.3.10 and 5.3.10(A)(B)

Delmar, *Firefighter's Handbook*, 3rd Edition, 1st Printing, page 259.

IFSTA, *Essentials of Fire Fighting and Fire Department Operations*, 5th Edition, 1st Printing, page 639–640.

Jones and Bartlett, NFPA, *Fundamentals of Fire Fighter Skills*, 2nd Edition, 1st Printing, page 467.

Answer: B

72. Reference: NFPA 1001, 5.3.10, 5.3.10(A)(B), 5.3.8 and 5.3.8(A)(B)

Delmar, *Firefighter's Handbook*, 3rd Edition, 1st Printing, page 261.

IFSTA, *Essentials of Fire Fighting and Fire Department Operations*, 5th Edition, 1st Printing, page 651.

Jones and Bartlett, NFPA, *Fundamentals of Fire Fighter Skills*, 2nd Edition, 1st Printing, page 468.

Answer: D

73. Reference: NFPA 1001, 5.3.10, 5.3.10(A)(B), 5.3.8 and 5.3.8(A)(B)

Delmar, *Firefighter's Handbook*, 3rd Edition, 1st Printing, page 261.

IFSTA, *Essentials of Fire Fighting and Fire Department Operations*, 5th Edition, 1st Printing, page 649.

Jones and Bartlett, NFPA, *Fundamentals of Fire Fighter Skills*, 2nd Edition, 1st Printing, page 481.

Answer: C

74. Reference: NFPA 1001, 5.3.10, 5.3.10(A)(B), 5.3.8 and 5.3.8(A)(B)
Delmar, *Firefighter's Handbook*, 3rd Edition, 1st Printing, pages 264–265.
IFSTA, *Essentials of Fire Fighting and Fire Department Operations*, 5th Edition, 1st Printing, page 688, Skill sheet 13-1–5.
Jones and Bartlett, NFPA, *Fundamentals of Fire Fighter Skills*, 2nd Edition, 1st Printing, pages 470–471.
Answer: D

75. Reference: NFPA 1001, 5.3.10, 5.3.10(A)(B), 5.3.8 and 5.3.8(A)(B)
Delmar, *Firefighter's Handbook*, 3rd Edition, 1st Printing, page 287.
IFSTA, *Essentials of Fire Fighting and Fire Department Operations*, 5th Edition, 1st Printing, page 789.
Jones and Bartlett, NFPA, *Fundamentals of Fire Fighter Skills*, 2nd Edition, 1st Printing, page 6.
Answer: C

76. Reference: NFPA 1001, 5.3.10, 5.3.10(A)(B), 5.3.15, and 5.3.15(A)(B)
Delmar, *Firefighter's Handbook*, 3rd Edition, 1st Printing, page 304.
IFSTA, *Essentials of Fire Fighting and Fire Department Operations*, 5th Edition, 1st Printing, pages 665–666.
Jones and Bartlett, NFPA, *Fundamentals of Fire Fighter Skills*, 2nd Edition, 1st Printing, page 487.
Answer: B

77. Reference: NFPA 1001, 5.3.10, 5.3.10(A), 5.3.8, 5.3.8(A), 5.3.7 and 5.3.7(A)
Delmar, *Firefighter's Handbook*, 3rd Edition, 1st Printing, page 325.
IFSTA, *Essentials of Fire Fighting and Fire Department Operations*, 5th Edition, 1st Printing, page 717.
Jones and Bartlett, NFPA, *Fundamentals of Fire Fighter Skills*, 2nd Edition, 1st Printing, page 619.
Answer: C

78. Reference: NFPA 1001, 5.3.10 and 5.3.10(A)
Delmar, *Firefighter's Handbook*, 3rd Edition, 1st Printing, page 256.
IFSTA, *Essentials of Fire Fighting and Fire Department Operations*, 5th Edition, 1st Printing, pages 663 and 724.
Jones and Bartlett, NFPA, *Fundamentals of Fire Fighter Skills*, 2nd Edition, 1st Printing, page 466.
Answer: A

79. Reference: NFPA 1001, 5.3.10, 5.3.10(A)
Delmar, *Firefighter's Handbook*, 3rd Edition, 1st Printing, page 395.
IFSTA, *Essentials of Fire Fighting and Fire Department Operations*, 5th Edition, 1st Printing, page 147.
Jones and Bartlett, NFPA, *Fundamentals of Fire Fighter Skills*, 2nd Edition, 1st Printing, pages 157–158.
Answer: D

80. Reference: NFPA 1001, 5.3.10, 5.3.10(A)
Delmar, *Firefighter's Handbook*, 3rd Edition, 1st Printing, pages 396–397.
IFSTA, *Essentials of Fire Fighting and Fire Department Operations*, 5th Edition,
1st Printing, page 148.
Jones and Bartlett, NFPA, *Fundamentals of Fire Fighter Skills*, 2nd Edition,
1st Printing, page 159.
Answer: B

81. Reference: NFPA 1001, 5.3.11 and 5.3.11(A)(B)
Delmar, *Firefighter's Handbook*, 3rd Edition, 1st Printing, page 669.
IFSTA, *Essentials of Fire Fighting and Fire Department Operations*, 5th Edition,
1st Printing, page 574.
Jones and Bartlett, NFPA, *Fundamentals of Fire Fighter Skills*, 2nd Edition,
1st Printing, page 413.
Answer: C

82. Reference: NFPA 1001, 5.3.11, 5.3.11(A), 5.3.12, 5.3.12(A), 5.3.13 and 5.3.13(A)
Delmar, *Firefighter's Handbook*, 3rd Edition, 1st Printing, page 99.
IFSTA, *Essentials of Fire Fighting and Fire Department Operations*, 5th Edition,
1st Printing, page 95.
Jones and Bartlett, NFPA, *Fundamentals of Fire Fighter Skills*, 2nd Edition,
1st Printing, pages 132–133.
Answer: B

83. Reference: NFPA 1001, 5.3.11, 5.3.11(A), 5.3.12, 5.3.12(A), 5.3.13 and 5.3.13(A)
Delmar, *Firefighter's Handbook*, 3rd Edition, 1st Printing, pages 97–99.
IFSTA, *Essentials of Fire Fighting and Fire Department Operations*, 5th Edition,
1st Printing, page 93.
Jones and Bartlett, NFPA, *Fundamentals of Fire Fighter Skills*, 2nd Edition,
1st Printing, page 132.
Answer: A

84. Reference: NFPA 1001, 5.3.11, 5.3.11(A), 5.3.12, 5.3.12(A), 5.3.13 and 5.3.13(A)
Delmar, *Firefighter's Handbook*, 3rd Edition, 1st Printing, page 89.
IFSTA, *Essentials of Fire Fighting and Fire Department Operations*, 5th Edition,
1st Printing, page 130.
Jones and Bartlett, NFPA, *Fundamentals of Fire Fighter Skills*, 2nd Edition,
1st Printing, page 133.
Answer: B

85. Reference: NFPA 1001, 5.3.11, 5.3.11(A), 5.3.12, 5.3.12(A), 5.3.13 and 5.3.13(A)
Delmar, *Firefighter's Handbook*, 3rd Edition, 1st Printing, pages 95–96.
IFSTA, *Essentials of Fire Fighting and Fire Department Operations*, 5th Edition,
1st Printing, page 104.
Jones and Bartlett, NFPA, *Fundamentals of Fire Fighter Skills*, 2nd Edition,
1st Printing, page 138.
Answer: B

86. Reference: NFPA 1001, 5.3.11, 5.3.11(A), 5.3.12, 5.3.12(A), 5.3.13 and 5.3.13(A)
Delmar, *Firefighter's Handbook*, 3rd Edition, 1st Printing, pages 96–99 and 103.
IFSTA, *Essentials of Fire Fighting and Fire Department Operations*, 5th Edition, 1st Printing, pages 107–109.
Jones and Bartlett, NFPA, *Fundamentals of Fire Fighter Skills*, 2nd Edition, 1st Printing, page 131.
Answer: A

87. Reference: NFPA 1001, 5.3.11, 5.3.11(A), 5.3.12, 5.3.12(A), 5.3.13 and 5.3.13(A)
Delmar, *Firefighter's Handbook*, 3rd Edition, 1st Printing, page 95.
IFSTA, *Essentials of Fire Fighting and Fire Department Operations*, 5th Edition, 1st Printing, page 97.
Jones and Bartlett, NFPA, *Fundamentals of Fire Fighter Skills*, 2nd Edition, 1st Printing, page 141.
Answer: B

88. Reference: NFPA 1001, 5.3.11, 5.3.11(A), 5.3.10, 5.3.10(A), 5.3.12, 5.3.12(A), 5.3.13 and 5.3.13(A)
Delmar, *Firefighter's Handbook*, 3rd Edition, 1st Printing, pages 99, 636–637.
IFSTA, *Essentials of Fire Fighting and Fire Department Operations*, 5th Edition, 1st Printing, page 118.
Jones and Bartlett, NFPA, *Fundamentals of Fire Fighter Skills*, 2nd Edition, 1st Printing, page 137.
Answer: C

89. Reference: NFPA 1001, 5.3.11, 5.3.11(A), 5.3.10, 5.3.10(A), 5.3.12, 5.3.12(A), 5.3.13 and 5.3.13(A)
Delmar, *Firefighter's Handbook*, 3rd Edition, 1st Printing, page 107.
IFSTA, *Essentials of Fire Fighting and Fire Department Operations*, 5th Edition, 1st Printing, page 119.
Jones and Bartlett, NFPA, *Fundamentals of Fire Fighter Skills*, 2nd Edition, 1st Printing, pages 137–138.
Answer: D

90. Reference: NFPA 1001, 5.3.11, 5.3.11(A), 5.3.12, 5.3.12(A), 5.3.13 and 5.3.13(A)
Delmar, *Firefighter's Handbook*, 3rd Edition, 1st Printing, page 631.
IFSTA, *Essentials of Fire Fighting and Fire Department Operations*, 5th Edition, 1st Printing, page 108.
Jones and Bartlett, NFPA, *Fundamentals of Fire Fighter Skills*, 2nd Edition, 1st Printing, page 131.
Answer: D

91. Reference: NFPA 1001, 5.3.11, 5.3.11(A), 5.3.12 and 5.3.12(A)
Delmar, *Firefighter's Handbook*, 3rd Edition, 1st Printing, page 631.
IFSTA, *Essentials of Fire Fighting and Fire Department Operations*, 5th Edition, 1st Printing, page 541.
Jones and Bartlett, NFPA, *Fundamentals of Fire Fighter Skills*, 2nd Edition, 1st Printing, page 405.
Answer: A

92. Reference: NFPA 1001, 5.3.11, 5.3.11(A), 5.3.12 and 5.3.12(A)
Delmar, *Firefighter's Handbook*, 3rd Edition, 1st Printing, pages 637–639.
IFSTA, *Essentials of Fire Fighting and Fire Department Operations*, 5th Edition,
1st Printing, pages 548, 556, and 560.
Jones and Bartlett, NFPA, *Fundamentals of Fire Fighter Skills*, 2nd Edition,
1st Printing, page 405.
Answer: D

93. Reference: NFPA 1001, 5.3.11, 5.3.11(A), 5.3.8, 5.3.8(A), 5.3.12, and 5.3.12(A)
Delmar, *Firefighter's Handbook*, 3rd Edition, 1st Printing, page 102.
IFSTA, *Essentials of Fire Fighting and Fire Department Operations*, 5th Edition,
1st Printing, page 802.
Jones and Bartlett, NFPA, *Fundamentals of Fire Fighter Skills*, 2nd Edition,
1st Printing, page 136.
Answer: C

94. Reference: NFPA 1001, 5.3.11, 5.3.11(A), 5.3.10, 5.3.10(A), 5.3.12, and 5.3.12(A)
Delmar, *Firefighter's Handbook*, 3rd Edition, 1st Printing, pages 101 and 112.
IFSTA, *Essentials of Fire Fighting and Fire Department Operations*, 5th Edition,
1st Printing, page 117.
Jones and Bartlett, NFPA, *Fundamentals of Fire Fighter Skills*, 2nd Edition,
1st Printing, page 140.
Answer: C

95. Reference: NFPA 1001, 5.3.11, 5.3.11(A), 5.3.12, and 5.3.12(A)
Delmar, *Firefighter's Handbook*, 3rd Edition, 1st Printing, pages 96 and 215.
IFSTA, *Essentials of Fire Fighting and Fire Department Operations*, 5th Edition,
1st Printing, page 110.
Jones and Bartlett, NFPA, *Fundamentals of Fire Fighter Skills*, 2nd Edition,
1st Printing, page 134.
Answer: B

96. Reference: NFPA 1001, 5.3.11 and 5.3.11(A)(B)
Delmar, *Firefighter's Handbook*, 3rd Edition, 1st Printing, pages 645–646.
IFSTA, *Essentials of Fire Fighting and Fire Department Operations*, 5th Edition,
1st Printing, page 575.
Jones and Bartlett, NFPA, *Fundamentals of Fire Fighter Skills*, 2nd Edition,
1st Printing, page 416.
Answer: B

97. Reference: NFPA 1001, 5.3.12 and 5.3.12(A)(B)
Delmar, *Firefighter's Handbook*, 3rd Edition, 1st Printing, page 665.
IFSTA, *Essentials of Fire Fighting and Fire Department Operations*, 5th Edition,
1st Printing, pages 558–559.
Jones and Bartlett, NFPA, *Fundamentals of Fire Fighter Skills*, 2nd Edition,
1st Printing, page 420.
Answer: B

98. Reference: NFPA 1001, 5.3.12, 5.3.12(A), 5.3.11 and 5.3.11(A)
Delmar, *Firefighter's Handbook*, 3rd Edition, 1st Printing, page 390.
IFSTA, *Essentials of Fire Fighting and Fire Department Operations*, 5th Edition, 1st Printing, page 141.
Jones and Bartlett, NFPA, *Fundamentals of Fire Fighter Skills*, 2nd Edition, 1st Printing, page 155.
Answer: B

99. Reference: NFPA 1001, 5.3.12 and 5.3.12(A)
Delmar, *Firefighter's Handbook*, 3rd Edition, 1st Printing, page 637.
IFSTA, *Essentials of Fire Fighting and Fire Department Operations*, 5th Edition, 1st Printing, page 124.
Jones and Bartlett, NFPA, *Fundamentals of Fire Fighter Skills*, 2nd Edition, 1st Printing, page 140.
Answer: A

100. Reference: NFPA 1001, 5.3.12, 5.3.12(A), 5.3.11 and 5.3.11(A)
Delmar, *Firefighter's Handbook*, 3rd Edition, 1st Printing, page 403.
IFSTA, *Essentials of Fire Fighting and Fire Department Operations*, 5th Edition, 1st Printing, page 158.
Jones and Bartlett, NFPA, *Fundamentals of Fire Fighter Skills*, 2nd Edition, 1st Printing, page 422.
Answer: C

101. Reference: NFPA 1001, 5.3.12 and 5.3.12(A)
Delmar, *Firefighter's Handbook*, 3rd Edition, 1st Printing, page 648.
IFSTA, *Essentials of Fire Fighting and Fire Department Operations*, 5th Edition, 1st Printing, page 556.
Jones and Bartlett, NFPA, *Fundamentals of Fire Fighter Skills*, 2nd Edition, 1st Printing, page 411.
Answer: B

102. Reference: NFPA 1001, 5.3.13 and 5.3.13(A)(B)
Delmar, *Firefighter's Handbook*, 3rd Edition, 1st Printing, page 743.
IFSTA, *Essentials of Fire Fighting and Fire Department Operations*, 5th Edition, 1st Printing, pages 869 and 881.
Jones and Bartlett, NFPA, *Fundamentals of Fire Fighter Skills*, 2nd Edition, 1st Printing, page 570.
Answer: D

103. Reference: NFPA 1001, 5.3.13 and 5.3.13(A)(B)
Delmar, *Firefighter's Handbook*, 3rd Edition, 1st Printing, page 743.
IFSTA, *Essentials of Fire Fighting and Fire Department Operations*, 5th Edition, 1st Printing, page 884.
Jones and Bartlett, NFPA, *Fundamentals of Fire Fighter Skills*, 2nd Edition, 1st Printing, page 576.
Answer: A

104. Reference: NFPA 1001, 5.3.13 and 5.3.13(A)(B)
Delmar, *Firefighter's Handbook*, 3rd Edition, 1st Printing, page 743.
IFSTA, *Essentials of Fire Fighting and Fire Department Operations*, 5th Edition, 1st Printing, page 869.
Jones and Bartlett, NFPA, *Fundamentals of Fire Fighter Skills*, 2nd Edition, 1st Printing, page 575.
Answer: B

105. Reference: NFPA 1001, 5.3.13, 5.3.13(A)(B) and 5.3.10(A)
Delmar, *Firefighter's Handbook*, 3rd Edition, 1st Printing, page 744.
IFSTA, *Essentials of Fire Fighting and Fire Department Operations*, 5th Edition, 1st Printing, page 882.
Jones and Bartlett, NFPA, *Fundamentals of Fire Fighter Skills*, 2nd Edition, 1st Printing, page 573.
Answer: D

106. Reference: NFPA 1001, 5.3.13, 5.3.13(A)(B), 5.3.10, and 5.3.10(A)
Delmar, *Firefighter's Handbook*, 3rd Edition, 1st Printing, page 743.
IFSTA, *Essentials of Fire Fighting and Fire Department Operations*, 5th Edition, 1st Printing, page 884.
Jones and Bartlett, NFPA, *Fundamentals of Fire Fighter Skills*, 2nd Edition, 1st Printing, page 577.
Answer: A

107. Reference: NFPA 1001, 5.3.14 and 5.3.14(A)(B)
Delmar, *Firefighter's Handbook*, 3rd Edition, 1st Printing, page 736.
IFSTA, *Essentials of Fire Fighting and Fire Department Operations*, 5th Edition, 1st Printing, page 869.
Jones and Bartlett, NFPA, *Fundamentals of Fire Fighter Skills*, 2nd Edition, 1st Printing, page 555.
Answer: C

108. Reference: NFPA 1001, 5.3.14 and 5.3.14(A)(B)
Delmar, *Firefighter's Handbook*, 3rd Edition, 1st Printing, page 728.
IFSTA, *Essentials of Fire Fighting and Fire Department Operations*, 5th Edition, 1st Printing, page 876.
Jones and Bartlett, NFPA, *Fundamentals of Fire Fighter Skills*, 2nd Edition, 1st Printing, page 570.
Answer: D

109. Reference: NFPA 1001, 5.3.14 and 5.3.14(A)(B)
Delmar, *Firefighter's Handbook*, 3rd Edition, 1st Printing, pages 736 and 743.
IFSTA, *Essentials of Fire Fighting and Fire Department Operations*, 5th Edition, 1st Printing, pages 869, 870, and 881.
Jones and Bartlett, NFPA, *Fundamentals of Fire Fighter Skills*, 2nd Edition, 1st Printing, pages 555 and 570.
Answer: C

110. Reference: NFPA 1001, 5.3.14 and 5.3.14(A)(B)

Delmar, *Firefighter's Handbook*, 3rd Edition, 1st Printing, pages 737–738, 740–741.

IFSTA, *Essentials of Fire Fighting and Fire Department Operations*, 5th Edition, 1st Printing, pages 878, 879, and 880.

Jones and Bartlett, NFPA, *Fundamentals of Fire Fighter Skills*, 2nd Edition, 1st Printing, page 565.

Answer: D

111. Reference: NFPA 1001, 5.3.14 and 5.3.14(A)

Delmar, *Firefighter's Handbook*, 3rd Edition, 1st Printing, pages 366–367.

IFSTA, *Essentials of Fire Fighting and Fire Department Operations*, 5th Edition, 1st Printing, pages 850–851.

Jones and Bartlett, NFPA, *Fundamentals of Fire Fighter Skills*, 2nd Edition, 1st Printing, pages 480 and 957.

Answer: A

112. Reference: NFPA 1001, 5.3.14 and 5.3.14(A)

Delmar, *Firefighter's Handbook*, 3rd Edition, 1st Printing, pages 728, 729.

IFSTA, *Essentials of Fire Fighting and Fire Department Operations*, 5th Edition, 1st Printing, page 876.

Jones and Bartlett, NFPA, *Fundamentals of Fire Fighter Skills*, 2nd Edition, 1st Printing, page 564–570.

Answer: A

113. Reference: NFPA 1001, 5.3.14 and 5.3.14(A)

Delmar, *Firefighter's Handbook*, 3rd Edition, 1st Printing, page 361.

IFSTA, *Essentials of Fire Fighting and Fire Department Operations*, 5th Edition, 1st Printing, page 845.

Jones and Bartlett, NFPA, *Fundamentals of Fire Fighter Skills*, 1st Edition, 1st Printing, page 954.

Answer: C

114. Reference: NFPA 5.3.14 and 5.3.14(A)(B)

Delmar, *Firefighter's Handbook*, 3rd Edition, 1st Printing, page 371.

IFSTA, *Essentials of Fire Fighting and Fire Department Operations*, 5th Edition, 1st Printing, page 860. Skill sheet 16-1–2.

Jones and Bartlett, NFPA, *Fundamentals of Fire Fighter Skills*, 2nd Edition, 1st Printing, page 561.

Answer: D

115. Reference: NFPA 1001, 5.3.14 and 5.3.14(A)(B)

Delmar, *Firefighter's Handbook*, 3rd Edition, 1st Printing, page 729.

IFSTA, *Essentials of Fire Fighting and Fire Department Operations*, 5th Edition, 1st Printing, page 849.

Jones and Bartlett, NFPA, *Fundamentals of Fire Fighter Skills*, 2nd Edition, 1st Printing, page 565.

Answer: B

116. Reference: NFPA 1001, 5.3.14 and 5.3.14(A)

Delmar, *Firefighter's Handbook*, 3rd Edition, 1st Printing, page 353.

IFSTA, *Essentials of Fire Fighting and Fire Department Operations*, 5th Edition, 1st Printing, page 829.

Jones and Bartlett, NFPA, *Fundamentals of Fire Fighter Skills*, 2nd Edition, 1st Printing, page 943.

Answer: B

117. Reference: NFPA 1001, 5.3.15 and 5.3.15(A)(B)

Delmar, *Firefighter's Handbook*, 3rd Edition, 1st Printing, page 311.

IFSTA, *Essentials of Fire Fighting and Fire Department Operations*, 5th Edition, 1st Printing, pages 664–665.

Jones and Bartlett, NFPA, *Fundamentals of Fire Fighter Skills*, 2nd Edition, 1st Printing, page 487.

Answer: B

118. Reference: NFPA 1001, 5.3.15 and 5.3.15(A)

Delmar, *Firefighter's Handbook*, 3rd Edition, 1st Printing, page 238.

IFSTA, *Essentials of Fire Fighting and Fire Department Operations*, 5th Edition, 1st Printing, pages 599–600.

Jones and Bartlett, NFPA, *Fundamentals of Fire Fighter Skills*, 2nd Edition, 1st Printing, page 450.

Answer: C

119. Reference: NFPA 1001, 5.3.16 and 5.3.16(A)(B)

Delmar, *Firefighter's Handbook*, 3rd Edition, 1st Printing, page 215.

IFSTA, *Essentials of Fire Fighting and Fire Department Operations*, 5th Edition, 1st Printing, page 246.

Jones and Bartlett, NFPA, *Fundamentals of Fire Fighter Skills*, 2nd Edition, 1st Printing, pages 183–184.

Answer: C

120. Reference: NFPA 1001, 5.3.16 and 5.3.16(A)(B)

Delmar, *Firefighter's Handbook*, 3rd Edition, 1st Printing, page 218.

IFSTA, *Essentials of Fire Fighting and Fire Department Operations*, 5th Edition, 1st Printing, pages 249–250.

Jones and Bartlett, NFPA, *Fundamentals of Fire Fighter Skills*, 2nd Edition, 1st Printing, pages 188–189.

Answer: B

121. Reference: NFPA 1001, 5.3.16 and 5.3.16(A)(B)

Delmar, *Firefighter's Handbook*, 3rd Edition, 1st Printing, page 215.

IFSTA, *Essentials of Fire Fighting and Fire Department Operations*, 5th Edition, 1st Printing, pages 110 and 249.

Jones and Bartlett, NFPA, *Fundamentals of Fire Fighter Skills*, 2nd Edition, 1st Printing, page 182.

Answer: A

122. Reference: NFPA 1001, 5.3.16 and 5.3.16(A)(B)
Delmar, *Firefighter's Handbook*, 3rd Edition, 1st Printing, page 216.
IFSTA, *Essentials of Fire Fighting and Fire Department Operations*, 5th Edition, 1st Printing, page 249.
Jones and Bartlett, NFPA, *Fundamentals of Fire Fighter Skills*, 2nd Edition, 1st Printing, pages 188–189.
Answer: C

123. Reference: NFPA 1001, 5.3.16 and 5.3.16(A)(B)
Delmar, *Firefighter's Handbook*, 3rd Edition, 1st Printing, page 216.
IFSTA, *Essentials of Fire Fighting and Fire Department Operations*, 5th Edition, 1st Printing, page 111.
Jones and Bartlett, NFPA, *Fundamentals of Fire Fighter Skills*, 2nd Edition, 1st Printing, page 183.
Answer: D

124. Reference: NFPA 1001, 5.3.16 and 5.3.16(A)(B)
Delmar, *Firefighter's Handbook*, 3rd Edition, 1st Printing, page 218.
IFSTA, *Essentials of Fire Fighting and Fire Department Operations*, 5th Edition, 1st Printing, page 241.
Jones and Bartlett, NFPA, *Fundamentals of Fire Fighter Skills*, 2nd Edition, 1st Printing, page 189.
Answer: B

125. Reference: NFPA 1001, 5.3.16 and 5.3.16(A)(B)
Delmar, *Firefighter's Handbook*, 3rd Edition, 1st Printing, page 221.
IFSTA, *Essentials of Fire Fighting and Fire Department Operations*, 5th Edition, 1st Printing, page 253.
Jones and Bartlett, NFPA, *Fundamentals of Fire Fighter Skills*, 2nd Edition, 1st Printing, page 193.
Answer: C

126. Reference: NFPA 1001, 5.3.16 and 5.3.16(A)(B)
Delmar, *Firefighter's Handbook*, 3rd Edition, 1st Printing, page 218.
IFSTA, *Essentials of Fire Fighting and Fire Department Operations*, 5th Edition, 1st Printing, page 243.
Jones and Bartlett, NFPA, *Fundamentals of Fire Fighter Skills*, 2nd Edition, 1st Printing, page 188.
Answer: D

127. Reference: NFPA 1001, 5.3.16 and 5.3.16(A)
Delmar, *Firefighter's Handbook*, 3rd Edition, 1st Printing, page 218.
IFSTA, *Essentials of Fire Fighting and Fire Department Operations*, 5th Edition, 1st Printing, page 254.
Jones and Bartlett, NFPA, *Fundamentals of Fire Fighter Skills*, 2nd Edition, 1st Printing, page 190.
Answer: A

128. Reference: NFPA 1001, 5.3.16 and 5.3.16(A)(B)
Delmar, *Firefighter's Handbook*, 3rd Edition, 1st Printing, pages 215–216, 230.
IFSTA, *Essentials of Fire Fighting and Fire Department Operations*, 5th Edition, 1st Printing, pages 237 and 243.
Jones and Bartlett, NFPA, *Fundamentals of Fire Fighter Skills*, 2nd Edition, 1st Printing, pages 188–189.
Answer: D

129. Reference: NFPA 1001, 5.3.16 and 5.3.16(A)(B)
Delmar, *Firefighter's Handbook*, 3rd Edition, 1st Printing, page 225.
IFSTA, *Essentials of Fire Fighting and Fire Department Operations*, 5th Edition, 1st Printing, page 251.
Jones and Bartlett, NFPA, *Fundamentals of Fire Fighter Skills*, 2nd Edition, 1st Printing, page 198.
Answer: C

130. Reference: NFPA 1001, 5.3.16 and 5.3.16(A)(B)
Delmar, *Firefighter's Handbook*, 3rd Edition, 1st Printing, page 214.
IFSTA, *Essentials of Fire Fighting and Fire Department Operations*, 5th Edition, 1st Printing, page 233.
Jones and Bartlett, NFPA, *Fundamentals of Fire Fighter Skills*, 2nd Edition, 1st Printing, page 181.
Answer: D

131. Reference: NFPA 1001, 5.3.16 and 5.3.16(A)(B)
Delmar, *Firefighter's Handbook*, 3rd Edition, 1st Printing, page 216.
IFSTA, *Essentials of Fire Fighting and Fire Department Operations*, 5th Edition, 1st Printing, page 243.
Jones and Bartlett, NFPA, *Fundamentals of Fire Fighter Skills*, 2nd Edition, 1st Printing, page 188.
Answer: B

132. Reference: NFPA 1001, 5.3.16 and 5.3.16(A)
Delmar, *Firefighter's Handbook*, 3rd Edition, 1st Printing, page 218.
IFSTA, *Essentials of Fire Fighting and Fire Department Operations*, 5th Edition, 1st Printing, page 243.
Jones and Bartlett, NFPA, *Fundamentals of Fire Fighter Skills*, 2nd Edition, 1st Printing, page 189.
Answer: D

133. Reference: NFPA 1001, 5.3.16, and 5.3.16(A)(B)
Delmar, *Firefighter's Handbook*, 3rd Edition, 1st Printing, page 216.
IFSTA, *Essentials of Fire Fighting and Fire Department Operations*, 5th Edition, 1st Printing, page 237.
Jones and Bartlett, NFPA, *Fundamentals of Fire Fighter Skills*, 2nd Edition, 1st Printing, page 183.
Answer: C

134. Reference: NFPA 1001, 5.3.18 and 5.3.18(A)(B)
Delmar, *Firefighter's Handbook*, 3rd Edition, 1st Printing, page 574.
IFSTA, *Essentials of Fire Fighting and Fire Department Operations*, 5th Edition, 1st Printing, page 369.
Jones and Bartlett, NFPA, *Fundamentals of Fire Fighter Skills*, 2nd Edition, 1st Printing, page 637.
Answer: B

135. Reference: NFPA 1001, 5.3.18 and 5.3.18(A)(B)
Delmar, *Firefighter's Handbook*, 3rd Edition, 1st Printing, page 699.
IFSTA, *Essentials of Fire Fighting and Fire Department Operations*, 5th Edition, 1st Printing, pages 778–779.
Jones and Bartlett, NFPA, *Fundamentals of Fire Fighter Skills*, 2nd Edition, 1st Printing, page 279.
Answer: A

136. Reference: NFPA 1001, 5.3.19 and 5.3.19(A)
Delmar, *Firefighter's Handbook*, 3rd Edition, 1st Printing, page 683.
IFSTA, *Essentials of Fire Fighting and Fire Department Operations*, 5th Edition, 1st Printing, page 800.
Jones and Bartlett, NFPA, *Fundamentals of Fire Fighter Skills*, 2nd Edition, 1st Printing, page 600.
Answer: C

137. Reference: NFPA 1001, 5.5.1 and 5.5.1(A)(B)
Delmar, *Firefighter's Handbook*, 3rd Edition, 1st Printing, page 201.
IFSTA, *Essentials of Fire Fighting and Fire Department Operations*, 5th Edition, 1st Printing, page 206.
Jones and Bartlett, NFPA, *Fundamentals of Fire Fighter Skills*, 2nd Edition, 1st Printing, page 64.
Answer: C

138. Reference: NFPA 1001, 5.5.1 and 5.5.1(A)(B)
Delmar, *Firefighter's Handbook*, 3rd Edition, 1st Printing, page 174.
IFSTA, *Essentials of Fire Fighting and Fire Department Operations*, 5th Edition, 1st Printing, page 206.
Jones and Bartlett, NFPA, *Fundamentals of Fire Fighter Skills*, 2nd Edition, 1st Printing, page 65.
Answer: B

139. Reference: NFPA 1001, 5.5.1 and 5.5.1(A)(B)
Delmar, *Firefighter's Handbook*, 3rd Edition, 1st Printing, pages 199, 201,202.
IFSTA, *Essentials of Fire Fighting and Fire Department Operations*, 5th Edition, 1st Printing, page 205.
Jones and Bartlett, NFPA, *Fundamentals of Fire Fighter Skills*, 2nd Edition, 1st Printing, page 64.
Answer: A

140. Reference: NFPA 1001, 5.5.1 and 5.5.1(A)(B)
Delmar, *Firefighter's Handbook*, 3rd Edition, 1st Printing, page 598.
IFSTA, *Essentials of Fire Fighting and Fire Department Operations*, 5th Edition, 1st Printing, page 411.
Jones and Bartlett, NFPA, *Fundamentals of Fire Fighter Skills*, 2nd Edition, 1st Printing, page 231.
Answer: C

141. Reference: NFPA 1001, 5.5.1 and 5.5.1(A)(B)
Delmar, *Firefighter's Handbook*, 3rd Edition, 1st Printing, page 731.
IFSTA, *Essentials of Fire Fighting and Fire Department Operations*, 5th Edition, 1st Printing, page 873.
Jones and Bartlett, NFPA, *Fundamentals of Fire Fighter Skills*, 2nd Edition, 1st Printing, page 568.
Answer: C

142. Reference: NFPA 1001, 5.5.1 and 5.5.1(A)
Delmar, *Firefighter's Handbook*, 3rd Edition, 1st Printing, page 424.
IFSTA, *Essentials of Fire Fighting and Fire Department Operations*, 5th Edition, 1st Printing, page 479.
Jones and Bartlett, NFPA, *Fundamentals of Fire Fighter Skills*, 2nd Edition, 1st Printing, page 333.
Answer: C

143. Reference: NFPA 1001, 5.5.1 and 5.5.1(A)(B)
Delmar, *Firefighter's Handbook*, 3rd Edition, 1st Printing, page 422.
IFSTA, *Essentials of Fire Fighting and Fire Department Operations*, 5th Edition, 1st Printing, page 480.
Jones and Bartlett, NFPA, *Fundamentals of Fire Fighter Skills*, 2nd Edition, 1st Printing, page 332.
Answer: C

144. Reference: NFPA 1001, 5.5.2 and 5.5.2(A)(B)
Delmar, *Firefighter's Handbook*, 3rd Edition, 1st Printing, page 258.
IFSTA, *Essentials of Fire Fighting and Fire Department Operations*, 5th Edition, 1st Printing, page 637.
Jones and Bartlett, NFPA, *Fundamentals of Fire Fighter Skills*, 2nd Edition, 1st Printing, pages 476–477.
Answer: C

145. Reference: NFPA 1001, 5.5.2, 5.5.2(A)(B), 5.3.15 and 5.3.15(A)(B)
Delmar, *Firefighter's Handbook*, 3rd Edition, 1st Printing, page 273.
IFSTA, *Essentials of Fire Fighting and Fire Department Operations*, 5th Edition, 1st Printing, page 656.
Jones and Bartlett, NFPA, *Fundamentals of Fire Fighter Skills*, 2nd Edition, 1st Printing, page 493.
Answer: D

146. Reference: NFPA 1001, 5.5.2, 5.5.2(A)(B), 5.3.10 and 5.3.10(A)(B)

Delmar, *Firefighter's Handbook*, 3rd Edition, 1st Printing, page 266.

IFSTA, *Essentials of Fire Fighting and Fire Department Operations*, 5th Edition, 1st Printing, page 653.

Jones and Bartlett, NFPA, *Fundamentals of Fire Fighter Skills*, 2nd Edition, 1st Printing, page 482.

Answer: B

147. Reference: NFPA 1001, 5.5.2, 5.5.2(A)(B), 5.3.15 and 5.3.15(A)(B)

Delmar, *Firefighter's Handbook*, 3rd Edition, 1st Printing, page 312.

IFSTA, *Essentials of Fire Fighting and Fire Department Operations*, 5th Edition, 1st Printing, page 666.

Jones and Bartlett, NFPA, *Fundamentals of Fire Fighter Skills*, 2nd Edition, 1st Printing, page 488.

Answer: C

148. Reference: NFPA 1001, 5.5.2 and 5.5.2(A)(B)

Delmar, *Firefighter's Handbook*, 3rd Edition, 1st Printing, pages 257–258.

IFSTA, *Essentials of Fire Fighting and Fire Department Operations*, 5th Edition, 1st Printing, page 634.

Jones and Bartlett, NFPA, *Fundamentals of Fire Fighter Skills*, 2nd Edition, 1st Printing, pages 466–476.

Answer: B

149. Reference: NFPA 1001, 5.5.2 and 5.5.2(A)(B)

Delmar, *Firefighter's Handbook*, 3rd Edition, 1st Printing, page 266.

IFSTA, *Essentials of Fire Fighting and Fire Department Operations*, 5th Edition, 1st Printing, page 653.

Jones and Bartlett, NFPA, *Fundamentals of Fire Fighter Skills*, 2nd Edition, 1st Printing, pages 482–483.

Answer: B

150. Reference: NFPA 1001, 5.5.2 and 5.5.2(A)(B)

Delmar, *Firefighter's Handbook*, 3rd Edition, 1st Printing, page 279.

IFSTA, *Essentials of Fire Fighting and Fire Department Operations*, 5th Edition, 1st Printing, pages 662–663.

Jones and Bartlett, NFPA, *Fundamentals of Fire Fighter Skills*, 2nd Edition, 1st Printing, page 499.

Answer: C

Don't forget to enter the information on your Personal Progress Plotter and answer the Yes and No question at the end of the Examination. This step is extremely important for the successful completion of the Systematic Approach to Examination Preparation!

APPENDIX B

Examination II-1 Answer Key

Directions
Follow these steps carefully for completing the feedback part of SAEP:

1. After calculating your score, look up the answers for the examination items you missed as well as those on which you guessed, even if you guessed correctly. If you are guessing, it means the answer is not perfectly clear. In this process, we are committed to making you as knowledgeable as possible.

2. Enter the number of missed and guessed examination items in the blanks on your Personal Progress Plotter.

3. Highlight the answer in the reference materials. Read the paragraph preceding and the paragraph following the one in which the correct answer is located. Enter the paragraph number and page number next to the guessed or missed examination item on your examination. Count any part of a paragraph at the beginning of the page as one paragraph until you reach the paragraph containing your highlighted answer. This step will help you locate and review your missed and guessed examination items later in the process. This step is essential to learning the material in context and by association. These learning techniques (context/association) are the very backbone of the SAEP approach.

4. Once you have completed the feedback step, you may proceed to the next examination.

1. Reference: NFPA 1001, 6.1.1 and 6.1.2
   Delmar, *Firefighter's Handbook*, 3rd Edition, 1st Printing, pages 39–40.
   IFSTA, *Essentials of Fire Fighting and Fire Department Operations*, 5th Edition, 1st Printing, page 792.
   Jones and Bartlett, NFPA, *Fundamentals of Fire Fighter Skills*, 2nd Edition, 1st Printing, page 119.
   Answer: B

2. Reference: NFPA 1001, 6.1.1 and 6.1.2
   Delmar, *Firefighter's Handbook*, 3rd Edition, 1st Printing, page 39.
   IFSTA, *Essentials of Fire Fighting and Fire Department Operations*, 5th Edition, 1st Printing, page 19.
   Jones and Bartlett, NFPA, *Fundamentals of Fire Fighter Skills*, 2nd Edition, 1st Printing, page 107.
   Answer: D

3. Reference: NFPA 1001, 6.1.1 and 6.1.2

Delmar, *Firefighter's Handbook*, 3rd Edition, 1st Printing, page 37.

IFSTA, *Essentials of Fire Fighting and Fire Department Operations*, 5th Edition, 1st Printing, page 33.

Jones and Bartlett, NFPA, *Fundamentals of Fire Fighter Skills*, 2nd Edition, 1st Printing, page 6.

Answer: B

4. Reference: NFPA 1001, 6.1.1

Delmar, *Firefighter's Handbook*, 3rd Edition, 1st Printing, page 37.

IFSTA, *Essentials of Fire Fighting and Fire Department Operations*, 5th Edition, 1st Printing, page 33.

Jones and Bartlett, NFPA, *Fundamentals of Fire Fighter Skills*, 2nd Edition, 1st Printing, page 6.

Answer: B

5. Reference: NFPA 1001, 6.1.1 and 6.1.2

Delmar, *Firefighter's Handbook*, 3rd Edition, 1st Printing, page 42.

IFSTA, *Essentials of Fire Fighting and Fire Department Operations*, 5th Edition, 1st Printing, page 36.

Jones and Bartlett, NFPA, *Fundamentals of Fire Fighter Skills*, 2nd Edition, 1st Printing, page 109.

Answer: D

6. Reference: NFPA 1001, 6.1.1 and 6.1.2

Delmar, *Firefighter's Handbook*, 3rd Edition, 1st Printing, pages 189, 824–825.

IFSTA, *Essentials of Fire Fighting and Fire Department Operations*, 5th Edition, 1st Printing, page 74.

Jones and Bartlett, NFPA, *Fundamentals of Fire Fighter Skills*, 2nd Edition, 1st Printing, page 32.

Answer: A

7. Reference: NFPA 1001, 6.1.1 and 6.1.2

Delmar, *Firefighter's Handbook*, 3rd Edition, 1st Printing, page 40.

IFSTA, *Essentials of Fire Fighting and Fire Department Operations*, 5th Edition, 1st Printing, page 35.

Jones and Bartlett, NFPA, *Fundamentals of Fire Fighter Skills*, 2nd Edition, 1st Printing, page 105.

Answer: B

8. Reference: NFPA 1001, 6.1.1 and 6.1.2

Delmar, *Firefighter's Handbook*, 3rd Edition, 1st Printing, page 41.

IFSTA, *Essentials of Fire Fighting and Fire Department Operations*, 5th Edition, 1st Printing, pages 35–36.

Jones and Bartlett, NFPA, *Fundamentals of Fire Fighter Skills*, 2nd Edition, 1st Printing, pages 107–108.

Answer: C

9. Reference: NFPA 1001, 6.2.1 and 6.2.1(A)(B)
Delmar, *Firefighter's Handbook*, 3rd Edition, 1st Printing, page 79.
IFSTA, *Essentials of Fire Fighting and Fire Department Operations*, 5th Edition, 1st Printing, page 949.
Jones and Bartlett, NFPA, *Fundamentals of Fire Fighter Skills*, 2nd Edition, 1st Printing, page 97.
Answer: A

10. Reference: NFPA 1001, 6.2.2, 6.2.2(A)(B), 6.3.2 and 6.3.2(A)(B)
Delmar, *Firefighter's Handbook*, 3rd Edition, 1st Printing, page 835.
IFSTA, *Essentials of Fire Fighting and Fire Department Operations*, 5th Edition, 1st Printing, page 948.
Jones and Bartlett, NFPA, *Fundamentals of Fire Fighter Skills*, 2nd Edition, 1st Printing, page 96.
Answer: D

11. Reference: NFPA 1001, 6.2.2 and 6.2.2(A)(B)
Delmar, *Firefighter's Handbook*, 3rd Edition, 1st Printing, page 75.
IFSTA, *Essentials of Fire Fighting and Fire Department Operations*, 5th Edition, 1st Printing, page 948.
Jones and Bartlett, NFPA, *Fundamentals of Fire Fighter Skills*, 2nd Edition, 1st Printing, page 96.
Answer: A

12. Reference: NFPA 1001, 6.2.2 and 6.2.2(A)(B)
Delmar, *Firefighter's Handbook*, 3rd Edition, 1st Printing, page 76.
IFSTA, *Essentials of Fire Fighting and Fire Department Operations*, 5th Edition, 1st Printing, page 946.
Jones and Bartlett, NFPA, *Fundamentals of Fire Fighter Skills*, 2nd Edition, 1st Printing, page 96.
Answer: D

13. Reference: NFPA 1001, 6.2.2 and 6.2.2(A)(B)
Delmar, *Firefighter's Handbook*, 3rd Edition, 1st Printing, page 72.
IFSTA, *Essentials of Fire Fighting and Fire Department Operations*, 5th Edition, 1st Printing, page 943.
Jones and Bartlett, NFPA, *Fundamentals of Fire Fighter Skills*, 2nd Edition, 1st Printing, page 91.
Answer: C

14. Reference: NFPA 1001, 6.2.2 and 6.2.2(A)(B)
Delmar, *Firefighter's Handbook*, 3rd Edition, 1st Printing, pages 72–73.
IFSTA, *Essentials of Fire Fighting and Fire Department Operations*, 5th Edition, 1st Printing, page 947.
Jones and Bartlett, NFPA, *Fundamentals of Fire Fighter Skills*, 2nd Edition, 1st Printing, page 92.
Answer: D

15. Reference: NFPA 1001, 6.2.2 and 6.2.2(A)(B)
Delmar, *Firefighter's Handbook*, 3rd Edition, 1st Printing, pages 65–66.
IFSTA, *Essentials of Fire Fighting and Fire Department Operations*, 5th Edition, 1st Printing, page 932.
Jones and Bartlett, NFPA, *Fundamentals of Fire Fighter Skills*, 2nd Edition, 1st Printing, page 84.
Answer: B

16. Reference: NFPA 1001, 6.2.2 and 6.2.2(A)(B)
Delmar, *Firefighter's Handbook*, 3rd Edition, 1st Printing, page 61.
IFSTA, *Essentials of Fire Fighting and Fire Department Operations*, 5th Edition, 1st Printing, page 938.
Jones and Bartlett, NFPA, *Fundamentals of Fire Fighter Skills*, 2nd Edition, 1st Printing, pages 83–85.
Answer: A

17. Reference: NFPA 1001, 6.2.2 and 6.2.2(A)(B)
Delmar, *Firefighter's Handbook*, 3rd Edition, 1st Printing, page 58.
IFSTA, *Essentials of Fire Fighting and Fire Department Operations*, 5th Edition, 1st Printing, page 934.
Jones and Bartlett, NFPA, *Fundamentals of Fire Fighter Skills*, 2nd Edition, 1st Printing, page 81.
Answer: A

18. Reference: NFPA 1001, 6.3.1 and 6.3.1(A)(B)
Delmar, *Firefighter's Handbook*, 3rd Edition, 1st Printing, page 341.
IFSTA, *Essentials of Fire Fighting and Fire Department Operations*, 5th Edition, 1st Printing, page 735.
Jones and Bartlett, NFPA, *Fundamentals of Fire Fighter Skills*, 2nd Edition, 1st Printing, page 190.
Answer: C

19. Reference: NFPA 1001, 6.3.1 and 6.3.1(A)(B)
Delmar, *Firefighter's Handbook*, 3rd Edition, 1st Printing, page 341.
IFSTA, *Essentials of Fire Fighting and Fire Department Operations*, 5th Edition, 1st Printing, page 742.
Jones and Bartlett, NFPA, *Fundamentals of Fire Fighter Skills*, 2nd Edition, 1st Printing, page 522.
Answer: B

20. References: NFPA 1001, 6.3.1, and 6.3.1(A)(B)
Delmar, *Firefighter's Handbook*, 3rd Edition, 1st Printing, page 344.
IFSTA, *Essentials of Fire Fighting and Fire Department Operations*, 5th Edition, 1st Printing, pages 744–745.
Jones and Bartlett, NFPA, *Fundamentals of Fire Fighter Skills*, 2nd Edition, 1st Printing, page 523.
Answer: C

21. Reference: NFPA 1001, 6.3.1 and 6.3.1(A)(B)
Delmar, *Firefighter's Handbook*, 3rd Edition, 1st Printing, pages 345–346.
IFSTA, *Essentials of Fire Fighting and Fire Department Operations*, 5th Edition,
1st Printing, page 747.
Jones and Bartlett, NFPA, *Fundamentals of Fire Fighter Skills*, 2nd Edition,
1st Printing, page 523.
Answer: C

22. Reference: NFPA 1001, 6.3.2 and 6.3.2(A)(B)
Delmar, *Firefighter's Handbook*, 3rd Edition, 1st Printing, page 406.
IFSTA, *Essentials of Fire Fighting and Fire Department Operations*, 5th Edition,
1st Printing, page 557.
Jones and Bartlett, NFPA, *Fundamentals of Fire Fighter Skills*, 2nd Edition,
1st Printing, page 165.
Answer: C

23. Reference: NFPA 1001, 6.3.2 and 6.3.2(A)(B)
Delmar, *Firefighter's Handbook*, 3rd Edition, 1st Printing, page 391.
IFSTA, *Essentials of Fire Fighting and Fire Department Operations*, 5th Edition,
1st Printing, pages 144 and 146.
Jones and Bartlett, NFPA, *Fundamentals of Fire Fighter Skills*, 2nd Edition,
1st Printing, page 154.
Answer: D

24. Reference: NFPA 1001, 6.3.2 and 6.3.2(A)(B)
Delmar, *Firefighter's Handbook*, 3rd Edition, 1st Printing, page 398.
IFSTA, *Essentials of Fire Fighting and Fire Department Operations*, 5th Edition,
1st Printing, page 149.
Jones and Bartlett, NFPA, *Fundamentals of Fire Fighter Skills*, 2nd Edition,
1st Printing, page 160.
Answer: D

25. Reference: NFPA 1001, 6.3.2 and 6.3.2(A)(B)
Delmar, *Firefighter's Handbook*, 3rd Edition, 1st Printing, page 403.
IFSTA, *Essentials of Fire Fighting and Fire Department Operations*, 5th Edition,
1st Printing, page 158.
Jones and Bartlett, NFPA, *Fundamentals of Fire Fighter Skills*, 2nd Edition,
1st Printing, page 169.
Answer: D

26. Reference: NFPA 1001, 6.3.2 and 6.3.2(A)(B)
Delmar, *Firefighter's Handbook*, 3rd Edition, 1st Printing, page 403.
IFSTA, *Essentials of Fire Fighting and Fire Department Operations*, 5th Edition,
1st Printing, page 158.
Jones and Bartlett, NFPA, *Fundamentals of Fire Fighter Skills*, 2nd Edition,
1st Printing, page 423.
Answer: C

27. Reference: NFPA 1001, 6.3.2 and 6.3.2(A)(B)

Delmar, *Firefighter's Handbook*, 3rd Edition, 1st Printing, page 597.

IFSTA, *Essentials of Fire Fighting and Fire Department Operations*, 5th Edition, 1st Printing, page 409.

Jones and Bartlett, NFPA, *Fundamentals of Fire Fighter Skills*, 2nd Edition, 1st Printing, page 294.

Answer: D

28. Reference: NFPA 1001, 6.3.2 and 6.3.2(A)(B)

Delmar, *Firefighter's Handbook*, 3rd Edition, 1st Printing, page 91.

IFSTA, *Essentials of Fire Fighting and Fire Department Operations*, 5th Edition, 1st Printing, page 92.

Jones and Bartlett, NFPA, *Fundamentals of Fire Fighter Skills*, 2nd Edition, 1st Printing, page 130.

Answer: D

29. Reference: NFPA 1001, 6.3.2 and 6.3.2(A)(B)

Delmar, *Firefighter's Handbook*, 3rd Edition, 1st Printing, pages 655–656.

IFSTA, *Essentials of Fire Fighting and Fire Department Operations*, 5th Edition, 1st Printing, page 560.

Jones and Bartlett, NFPA, *Fundamentals of Fire Fighter Skills*, 2nd Edition, 1st Printing, page 411.

Answer: C

30. Reference: NFPA 1001, 6.3.2 and 6.3.2(A)(B)

Delmar, *Firefighter's Handbook*, 3rd Edition, 1st Printing, pages 655–656.

IFSTA, *Essentials of Fire Fighting and Fire Department Operations,* 5th Edition, 1st Printing, pages 560–561.

Jones and Bartlett, NFPA, *Fundamentals of Fire Fighter Skills*, 2nd Edition, 1st Printing, page 420.

Answer: B

31. Reference: NFPA 1001, 6.3.2 and 6.3.2(A)(B)

Delmar, *Firefighter's Handbook*, 3rd Edition, 1st Printing, page 656.

IFSTA, *Essentials of Fire Fighting and Fire Department Operations*, 5th Edition, 1st Printing, page 560.

Jones and Bartlett, NFPA, *Fundamentals of Fire Fighter Skills*, 2nd Edition, 1st Printing, page 424.

Answer: B

32. Reference: NFPA 1001, 6.3.2 and 6.3.2(A)(B)

Delmar, *Firefighter's Handbook*, 3rd Edition, 1st Printing, page 656.

IFSTA, *Essentials of Fire Fighting and Fire Department Operations*, 5th Edition, 1st Printing, page 544.

Jones and Bartlett, NFPA, *Fundamentals of Fire Fighter Skills*, 2nd Edition, 1st Printing, page 424.

Answer: B

33. Reference: NFPA 1001, 6.3.2 and 6.3.2(A)(B)

Delmar, *Firefighter's Handbook*, 3rd Edition, 1st Printing, pages 377 and 705.

IFSTA, *Essentials of Fire Fighting and Fire Department Operations*, 5th Edition, 1st Printing, page 793.

Jones and Bartlett, NFPA, *Fundamentals of Fire Fighter Skills*, 2nd Edition, 1st Printing, page 514.

Answer: B

34. Reference: NFPA 1001, 6.3.2 and 6.3.2(A)(B)

Delmar, *Firefighter's Handbook*, 3rd Edition, 1st Printing, pages 395–396.

IFSTA, *Essentials of Fire Fighting and Fire Department Operations*, 5th Edition, 1st Printing, page 148.

Jones and Bartlett, NFPA, *Fundamentals of Fire Fighter Skills*, 2nd Edition, 1st Printing, page 159.

Answer: C

35. Reference: NFPA 1001, 6.3.2 and 6.3.2(A)(B)

Delmar, *Firefighter's Handbook*, 3rd Edition, 1st Printing, page 362.

IFSTA, *Essentials of Fire Fighting and Fire Department Operations*, 5th Edition, 1st Printing, page 856.

Jones and Bartlett, NFPA, *Fundamentals of Fire Fighter Skills*, 2nd Edition, 1st Printing, page 960.

Answer: B

36. References: NFPA 1001, 6.3.2 and 6.3.2(A)(B)

Delmar, *Firefighter's Handbook*, 3rd Edition, 1st Printing, page 391.

IFSTA, *Essentials of Fire Fighting and Fire Department Operations*, 5th Edition, 1st Printing, page 144.

Jones and Bartlett, NFPA, *Fundamentals of Fire Fighter Skills*, 2nd Edition, 1st Printing, page 154.

Answer: C

37. Reference: NFPA 1001, 6.3.2 and 6.3.2(A)(B)

Delmar, *Firefighter's Handbook*, 3rd Edition, 1st Printing, page 836.

IFSTA, *Essentials of Fire Fighting and Fire Department Operations*, 5th Edition, 1st Printing, page 71.

Jones and Bartlett, NFPA, *Fundamentals of Fire Fighter Skills*, 2nd Edition, 1st Printing, page 535.

Answer: B

38. Reference: NFPA 1001, 6.3.2 and 6.3.2(A)(B)

Delmar, *Firefighter's Handbook*, 3rd Edition, 1st Printing, page 328.

IFSTA, *Essentials of Fire Fighting and Fire Department Operations*, 5th Edition, 1st Printing, page 731.

Jones and Bartlett, NFPA, *Fundamentals of Fire Fighter Skills*, 2nd Edition, 1st Printing, page 517.

Answer: B

39. Reference: NFPA 1001, 6.3.3 and 6.3.3(A)(B)

Delmar, *Firefighter's Handbook*, 3rd Edition, 1st Printing, pages 891 and 1001.

IFSTA, *Essentials of Fire Fighting and Fire Department Operations*, 5th Edition, 1st Printing, pages 774, 775, and 780.

Jones and Bartlett, NFPA, *Fundamentals of Fire Fighter Skills*, 2nd Edition, 1st Printing, page 635.

Answer: B

40. Reference: NFPA 1001, 6.3.3 and 6.3.3(A)(B)

Delmar, *Firefighter's Handbook*, 3rd Edition, 1st Printing, page 890.

IFSTA, *Essentials of Fire Fighting and Fire Department Operations*, 5th Edition, 1st Printing, page 774.

Jones and Bartlett, NFPA, *Fundamentals of Fire Fighter Skills*, 2nd Edition, 1st Printing, page 635.

Answer: B

41. Reference: NFPA 1001, 6.3.3 and 6.3.3(A)(B)

Delmar, *Firefighter's Handbook*, 3rd Edition, 1st Printing, page 908.

IFSTA, *Essentials of Fire Fighting and Fire Department Operations*, 5th Edition, 1st Printing, page 779.

Jones and Bartlett, NFPA, *Fundamentals of Fire Fighter Skills*, 2nd Edition, 1st Printing, page 141.

Answer: C

42. Reference: NFPA 1001, 6.3.4 and 6.3.4(A)(B)

Delmar, *Firefighter's Handbook*, 3rd Edition, 1st Printing, pages 747–748.

IFSTA, *Essentials of Fire Fighting and Fire Department Operations*, 5th Edition, 1st Printing, page 918.

Jones and Bartlett, NFPA, *Fundamentals of Fire Fighter Skills*, 2nd Edition, 1st Printing, pages 977 and 984.

Answer: B

43. Reference: NFPA 1001, 6.3.4 and 6.3.4(A)(B)

Delmar, *Firefighter's Handbook*, 3rd Edition, 1st Printing, page 747.

IFSTA, *Essentials of Fire Fighting and Fire Department Operations*, 5th Edition, 1st Printing, page 918.

Jones and Bartlett, NFPA, *Fundamentals of Fire Fighter Skills*, 2nd Edition, 1st Printing, page 977.

Answer: D

44. References: NFPA 1001, 6.3.4, and 6.3.4(A)(B)

Delmar, *Firefighter's Handbook*, 2nd Edition, 1st Printing, pages 747–748.

IFSTA, *Essentials of Fire Fighting and Fire Department Operations*, 5th Edition, 1st Printing, page 918.

Jones and Bartlett, NFPA, *Fundamentals of Fire Fighter Skills*, 2nd Edition, 1st Printing, page 977.

Answer: D

45. Reference: NFPA 1001, 6.4.1 and 6.4.1(A)(B)

Delmar, *Firefighter's Handbook*, 3rd Edition, 1st Printing, page 553.

IFSTA, *Essentials of Fire Fighting and Fire Department Operations*, 5th Edition, 1st Printing, page 353.

Jones and Bartlett, NFPA, *Fundamentals of Fire Fighter Skills*, 2nd Edition, 1st Printing, page 725.

Answer: B

46. Reference: NFPA 1001, 6.4.1 and 6.4.1(A)(B)

Delmar, *Firefighter's Handbook*, 3rd Edition, 1st Printing, page 553.

IFSTA, *Essentials of Fire Fighting and Fire Department Operations*, 5th Edition, 1st Printing, page 353.

Jones and Bartlett, NFPA, *Fundamentals of Fire Fighter Skills*, 2nd Edition, 1st Printing, page 733.

Answer: B

47. Reference: NFPA 1001, 6.4.1 and 6.4.1(A)(B)

Delmar, *Firefighter's Handbook*, 3rd Edition, 1st Printing, page 555.

IFSTA, *Essentials of Fire Fighting and Fire Department Operations*, 5th Edition, 1st Printing, page 358.

Jones and Bartlett, NFPA, *Fundamentals of Fire Fighter Skills*, 2nd Edition, 1st Printing, page 735.

Answer: B

48. Reference: NFPA 1001, 6.4.1, 6.4.1(A)(B), 6.4.2, and 6.4.2(A)(B)

Delmar, *Firefighter's Handbook*, 3rd Edition, 1st Printing, page 550.

IFSTA, *Essentials of Fire Fighting and Fire Department Operations*, 5th Edition, 1st Printing, page 345.

Jones and Bartlett, NFPA, *Fundamentals of Fire Fighter Skills*, 2nd Edition, 1st Printing, page 731.

Answer: D

49. Reference: NFPA 1001, 6.4.1 and 6.4.1(A)(B)

Delmar, *Firefighter's Handbook*, 3rd Edition, 1st Printing, page 557.

IFSTA, *Essentials of Fire Fighting and Fire Department Operations*, 5th Edition, page 361.

Jones and Bartlett, NFPA, *Fundamentals of Fire Fighter Skills*, 2nd Edition, 1st Printing, page 741.

Answer: C

50. Reference: NFPA 1001, 6.4.2 and 6.4.2(A)(B)

Delmar, *Firefighter's Handbook*, 3rd Edition, 1st Printing, pages 570–571.

IFSTA, *Essentials of Fire Fighting and Fire Department Operations*, 5th Edition, 1st Printing, page 365.

Jones and Bartlett, NFPA, *Fundamentals of Fire Fighter Skills*, 2nd Edition, 1st Printing, page 763.

Answer: A

51. Reference: NFPA 1001, 6.4.2 and 6.4.2(A)(B)

Delmar, *Firefighter's Handbook*, 3rd Edition, 1st Printing, page 548.

IFSTA, *Essentials of Fire Fighting and Fire Department Operations*, 5th Edition, 1st Printing, page 335.

Jones and Bartlett, NFPA, *Fundamentals of Fire Fighter Skills*, 2nd Edition, 1st Printing, page 295.

Answer: B

52. Reference: NFPA 1001, 6.4.2 and 6.4.2(A)(B)

Delmar, *Firefighter's Handbook*, 3rd Edition, 1st Printing, pages 536–537.

IFSTA, *Essentials of Fire Fighting and Fire Department Operations*, 5th Edition, 1st Printing, page 380.

Jones and Bartlett, NFPA, *Fundamentals of Fire Fighter Skills*, 2nd Edition, 1st Printing, page 385.

Answer: B

53. Reference: NFPA 1001, 6.5.1 and 6.5.1(A)(B)

Delmar, *Firefighter's Handbook*, 3rd Edition, 1st Printing, pages 763–764.

IFSTA, *Essentials of Fire Fighting and Fire Department Operations,* 5th Edition, 1st Printing, pages 971–972.

Jones and Bartlett, NFPA, *Fundamentals of Fire Fighter Skills*, 2nd Edition, 1st Printing, pages 650–651.

Answer: B

54. References: NFPA 1001, 6.5.1 and 6.5.1(A)(B)

Delmar, *Firefighter's Handbook*, 3rd Edition, 1st Printing, page 395.

IFSTA, *Essentials of Fire Fighting and Fire Department Operations,* 5th Edition, 1st Printing, pages 147.

Jones and Bartlett, NFPA, *Fundamentals of Fire Fighter Skills*, 2nd Edition, 1st Printing, page 648.

Answer: B

55. Reference: NFPA 1001, 6.5.1 and 6.5.1(A)(B)

Delmar, *Firefighter's Handbook*, 3rd Edition, 1st Printing, page 756.

IFSTA, *Essentials of Fire Fighting and Fire Department Operations*, 5th Edition, 1st Printing, page 961.

Jones and Bartlett, NFPA, *Fundamentals of Fire Fighter Skills*, 2nd Edition, 1st Printing, pages 643–644.

Answer: C

56. Reference: NFPA 1001, 6.5.1 and 6.5.1(A)

Delmar, *Firefighter's Handbook*, 3rd Edition, 1st Printing, page 782.

IFSTA, *Essentials of Fire Fighting and Fire Department Operations*, 5th Edition, 1st Printing, page 966.

Jones and Bartlett, NFPA, *Fundamentals of Fire Fighter Skills*, 2nd Edition, 1st Printing, page 644.

Answer: C

57. Reference: NFPA 1001, 6.5.1 and 6.5.1(A)(B)
Delmar, *Firefighter's Handbook*, 3rd Edition, 1st Printing, page 775.
IFSTA, *Essentials of Fire Fighting and Fire Department Operations*, 5th Edition, page 970.
Jones and Bartlett, NFPA, *Fundamentals of Fire Fighter Skills*, 2nd Edition, page 919.
Answer: C

58. Reference: NFPA 1001, 6.5.1 and 6.5.1(A)(B)
Delmar, *Firefighter's Handbook*, 2nd Edition, 1st Printing, page 777.
IFSTA, *Essentials of Fire Fighting and Fire Department Operations*, 5th Edition, 1st Printing, page 982.
Jones and Bartlett, NFPA, *Fundamentals of Fire Fighter Skills*, 2nd Edition, page 919.
Answer: C

59. Reference: NFPA 1001, 6.5.1 and 6.5.1(A)(B)
Delmar, *Firefighter's Handbook*, 3rd Edition, 1st Printing, page 757.
IFSTA, *Essentials of Fire Fighting and Fire Department Operations*, 5th Edition, 1st Printing, page 968.
Jones and Bartlett, NFPA, *Fundamentals of Fire Fighter Skills*, 2nd Edition, 1st Printing, page 645.
Answer: C

60. Reference: NFPA 1001, 6.5.2 and 6.5.2(A)
Delmar, *Firefighter's Handbook*, 3rd Edition, 1st Printing, page 754.
IFSTA, *Essentials of Fire Fighting and Fire Department Operations*, 5th Edition, 1st Printing, page 976.
Jones and Bartlett, NFPA, *Fundamentals of Fire Fighter Skills*, 2nd Edition, 1st Printing, page 918.
Answer: A

61. Reference: NFPA 1001, 6.5.2 and 6.5.2(A)(B)
Delmar, *Firefighter's Handbook*, 3rd Edition, 1st Printing, page 780
IFSTA, *Essentials of Fire Fighting and Fire Department Operations*, 5th Edition, 1st Printing, page 987.
Jones and Bartlett, NFPA, *Fundamentals of Fire Fighter Skills*, 2nd Edition, 1st Printing, page 919.
Answer: B

62. Reference: NFPA 1001, 6.5.2 and 6.5.2(A)(B)
Delmar, *Firefighter's Handbook*, 3rd Edition, 1st Printing, page 777.
IFSTA, *Essentials of Fire Fighting and Fire Department Operations*, 5th Edition, 1st Printing, page 985.
Jones and Bartlett, NFPA, *Fundamentals of Fire Fighter Skills*, 2nd Edition, 1st Printing, page 929.
Answer: D

63. Reference: NFPA 1001, 6.5.3 and 6.5.3(A)(B)

Delmar, *Firefighter's Handbook*, 3rd Edition, 1st Printing, page 353.

IFSTA, *Essentials of Fire Fighting and Fire Department Operations*, 5th Edition, 1st Printing, page 826.

Jones and Bartlett, NFPA, *Fundamentals of Fire Fighter Skills*, 2nd Edition, 1st Printing, page 940.

Answer: D

64. Reference: NFPA 1001, 6.5.3 and 6.5.3(A)(B)

Delmar, *Firefighter's Handbook*, 3rd Edition, 1st Printing, page 357.

IFSTA, *Essentials of Fire Fighting and Fire Department Operations*, 5th Edition, 1st Printing, pages 842–845, and 854.

Jones and Bartlett, NFPA, *Fundamentals of Fire Fighter Skills*, 2nd Edition, 1st Printing, page 950.

Answer: C

65. Reference: NFPA 1001, 6.5.3 and 6.5.3(A)(B)

Delmar, *Firefighter's Handbook*, 3rd Edition, 1st Printing, page 364.

IFSTA, *Essentials of Fire Fighting and Fire Department Operations*, 5th Edition, 1st Printing, page 854.

Jones and Bartlett, NFPA, *Fundamentals of Fire Fighter Skills*, 2nd Edition, 1st Printing, page 959.

Answer: C

66. Reference: NFPA 1001, 6.5.3 and 6.5.3(A)(B)

Delmar, *Firefighter's Handbook*, 3rd Edition, 1st Printing, page 366.

IFSTA, *Essentials of Fire Fighting and Fire Department Operations*, 5th Edition, 1st Printing, pages 850–851.

Jones and Bartlett, NFPA, *Fundamentals of Fire Fighter Skills*, 2nd Edition, 1st Printing, page 957.

Answer: C

67. Reference: NFPA 1001, 6.5.3 and 6.5.3(A)(B)

Delmar, *Firefighter's Handbook*, 3rd Edition, 1st Printing, page 367.

IFSTA, *Essentials of Fire Fighting and Fire Department Operations*, 5th Edition, 1st Printing, page 848.

Jones and Bartlett, NFPA, *Fundamentals of Fire Fighter Skills*, 2nd Edition, 1st Printing, page 954.

Answer: D

68. Reference: NFPA 1001, 6.5.3 and 6.5.3(A)(B)

Delmar, *Firefighter's Handbook*, 3rd Edition, 1st Printing, page 366.

IFSTA, *Essentials of Fire Fighting and Fire Department Operations*, 5th Edition, 1st Printing, page 855.

Jones and Bartlett, NFPA, *Fundamentals of Fire Fighter Skills*, 2nd Edition, 1st Printing, page 960.

Answer: D

69. Reference: NFPA 1001, 6.5.3 and 6.5.3(A)(B)

Delmar, *Firefighter's Handbook*, 3rd Edition, 1st Printing, page 361.

IFSTA, *Essentials of Fire Fighting and Fire Department Operations*, 5th Edition, 1st Printing, page 851.

Jones and Bartlett, NFPA, *Fundamentals of Fire Fighter Skills*, 2nd Edition, 1st Printing, page 956.

Answer: B

70. Reference: NFPA 1001, 6.5.3 and 6.5.3(A)(B)

Delmar, *Firefighter's Handbook*, 3rd Edition, 1st Printing, page 782.

IFSTA, *Essentials of Fire Fighting and Fire Department Operations*, 5th Edition, 1st Printing, page 970.

Jones and Bartlett, NFPA, *Fundamentals of Fire Fighter Skills*, 2nd Edition, 1st Printing, page 645.

Answer: D

71. Reference: NFPA 1001, 6.5.3 and 6.5.3(A)(B)

Delmar, *Firefighter's Handbook*, 3rd Edition, 1st Printing, page 367.

IFSTA, *Essentials of Fire Fighting and Fire Department Operations*, 5th Edition, 1st Printing, page 601.

Jones and Bartlett, NFPA, *Fundamentals of Fire Fighter Skills*, 2nd Edition, 1st Printing, page 954.

Answer: D

72. Reference: NFPA 1001, 6.5.3 and 6.5.3(A)(B)

Delmar, *Firefighter's Handbook*, 3rd Edition, 1st Printing, pages 324–325.

IFSTA, *Essentials of Fire Fighting and Fire Department Operations*, 5th Edition, 1st Printing, page 601.

Jones and Bartlett, NFPA, *Fundamentals of Fire Fighter Skills*, 2nd Edition, 1st Printing, page 954.

Answer: A

73. Reference: NFPA 1001, 6.5.3 and 6.5.3(A)(B)

Delmar, *Firefighter's Handbook*, 3rd Edition, 1st Printing, page 239.

IFSTA, *Essentials of Fire Fighting and Fire Department Operations*, 5th Edition, 1st Printing, page 606.

Jones and Bartlett, NFPA, *Fundamentals of Fire Fighter Skills*, 2nd Edition, 1st Printing, page 453.

Answer: C

74. Reference: NFPA 1001, 6.5.3 and 6.5.3(A)(B)

Delmar, *Firefighter's Handbook*, 3rd Edition, 1st Printing, pages 237–238.

IFSTA, *Essentials of Fire Fighting and Fire Department Operations*, 5th Edition, 1st Printing, page 599.

Jones and Bartlett, NFPA, *Fundamentals of Fire Fighter Skills*, 2nd Edition, 1st Printing, page 450.

Answer: D

75. Reference: NFPA 1001, 6.5.3 and 6.5.3(A)(B)

Delmar, *Firefighter's Handbook*, 3rd Edition, 1st Printing, page 238.

IFSTA, *Essentials of Fire Fighting and Fire Department Operations*, 5th Edition, 1st Printing, page 608.

Jones and Bartlett, NFPA, *Fundamentals of Fire Fighter Skills*, 2nd Edition, 1st Printing, page 452.

Answer: B

76. Reference: NFPA 1001, 6.5.3 and 6.5.3(A)(B)

Delmar, *Firefighter's Handbook*, 3rd Edition, 1st Printing, page 237.

IFSTA, *Essentials of Fire Fighting and Fire Department Operations*, 5th Edition, 1st Printing, page 595.

Jones and Bartlett, NFPA, *Fundamentals of Fire Fighter Skills*, 2nd Edition, 1st Printing, pages 449–450.

Answer: A

77. Reference: NFPA 1001, 6.5.3 and 6.5.3(A)(B)

Delmar, *Firefighter's Handbook*, 3rd Edition, 1st Printing, page 237.

IFSTA, *Essentials of Fire Fighting and Fire Department Operations*, 5th Edition, 1st Printing, page 594.

Jones and Bartlett, NFPA, *Fundamentals of Fire Fighter Skills*, 2nd Edition, 1st Printing, page 449.

Answer: D

78. Reference: NFPA 1001, 6.5.3 and 6.5.3(A)(B)

Delmar, *Firefighter's Handbook*, 3rd Edition, 1st Printing, page 238.

IFSTA, *Essentials of Fire Fighting and Fire Department Operations*, 5th Edition, 1st Printing, pages 599–600.

Jones and Bartlett, NFPA, *Fundamentals of Fire Fighter Skills*, 2nd Edition, 1st Printing, page 450.

Answer: D

79. Reference: NFPA 1001, 6.5.3 and 6.5.3(A)(B)

Delmar, *Firefighter's Handbook*, 3rd Edition, 1st Printing, page 238.

IFSTA, *Essentials of Fire Fighting and Fire Department Operations*, 5th Edition, 1st Printing, page 599.

Jones and Bartlett, NFPA, *Fundamentals of Fire Fighter Skills*, 2nd Edition, 1st Printing, page 450.

Answer: C

80. Reference: NFPA 1001, 6.5.3 and 6.5.3(A)(B)

Delmar, *Firefighter's Handbook*, 3rd Edition, 1st Printing, pages 237–238.

IFSTA, *Essentials of Fire Fighting and Fire Department Operations*, 5th Edition, 1st Printing, page 599.

Jones and Bartlett, NFPA, *Fundamentals of Fire Fighter Skills*, 2nd Edition, 1st Printing, page 450.

Answer: A

81. Reference: NFPA 1001, 6.5.3 and 6.5.3(A)(B)

Delmar, *Firefighter's Handbook*, 3rd Edition, 1st Printing, page 238.

IFSTA, *Essentials of Fire Fighting and Fire Department Operations*, 5th Edition, 1st Printing, page 606.

Jones and Bartlett, NFPA, *Fundamentals of Fire Fighter Skills*, 2nd Edition, 1st Printing, page 452.

Answer: A

82. Reference: NFPA 1001, 6.5.3 and 6.5.3(A)(B)

Delmar, *Firefighter's Handbook*, 3rd Edition, 1st Printing, pages 237–238.

IFSTA, *Essentials of Fire Fighting and Fire Department Operations*, 5th Edition, 1st Printing, page 599.

Jones and Bartlett, NFPA, *Fundamentals of Fire Fighter Skills*, 2nd Edition, 1st Printing, page 450.

Answer: B

83. Reference: NFPA 1001, 6.5.3 and 6.5.3(A)(B)

Delmar, *Firefighter's Handbook*, 3rd Edition, 1st Printing, page 365.

IFSTA, *Essentials of Fire Fighting and Fire Department Operations*, 5th Edition, 1st Printing, page 855.

Jones and Bartlett, NFPA, *Fundamentals of Fire Fighter Skills*, 2nd Edition, 1st Printing, pages 960–961.

Answer: A

84. Reference: NFPA 1001, 6.5.3 and 6.5.3(A)

Delmar, *Firefighter's Handbook*, 3rd Edition, 1st Printing, page 364.

IFSTA, *Essentials of Fire Fighting and Fire Department Operations*, 5th Edition, 1st Printing, page 854.

Jones and Bartlett, NFPA, *Fundamentals of Fire Fighter Skills*, 2nd Edition, 1st Printing, page 959.

Answer: C

85. Reference: NFPA 1001, 6.5.3 and 6.5.3(A)(B)

Delmar, *Firefighter's Handbook*, 3rd Edition, 1st Printing, page 367.

IFSTA, *Essentials of Fire Fighting and Fire Department Operations*, 5th Edition, 1st Printing, page 848.

Jones and Bartlett, NFPA, *Fundamentals of Fire Fighter Skills*, 2nd Edition, 1st Printing, page 954.

Answer: C

86. Reference: NFPA 1001, 6.5.3 and 6.5.3(A)(B)

Delmar, *Firefighter's Handbook*, 3rd Edition, 1st Printing, page 363.

IFSTA, *Essentials of Fire Fighting and Fire Department Operations*, 5th Edition, 1st Printing, page 853.

Jones and Bartlett, NFPA, *Fundamentals of Fire Fighter Skills*, 2nd Edition, 1st Printing, page 958.

Answer: D

87. Reference: NFPA 1001, 6.5.3 and 6.5.3(A)(B)

Delmar, *Firefighter's Handbook*, 3rd Edition, 1st Printing, page 239.

IFSTA, *Essentials of Fire Fighting and Fire Department Operations*, 5th Edition, 1st Printing, page 606.

Jones and Bartlett, NFPA, *Fundamentals of Fire Fighter Skills*, 2nd Edition, 1st Printing, page 453.

Answer: D

88. Reference: NFPA 1001, 6.5.3 and 6.5.3(A)(B)

Delmar, *Firefighter's Handbook*, 3rd Edition, 1st Printing, page 239.

IFSTA, *Essentials of Fire Fighting and Fire Department Operations*, 5th Edition, 1st Printing, pages 605–606.

Jones and Bartlett, NFPA, *Fundamentals of Fire Fighter Skills*, 2nd Edition, 1st Printing, pages 452–453.

Answer: D

89. Reference: NFPA 1001, 6.5.3 and 6.5.3(A)(B)

Delmar, *Firefighter's Handbook*, 2nd Edition, 1st Printing, page 247.

IFSTA, *Essentials of Fire Fighting and Fire Department Operations*, 5th Edition, 1st Printing, pages 604, 608–609.

Jones and Bartlett, NFPA, *Fundamentals of Fire Fighter Skills*, 2nd Edition, 1st Printing, pages 455–457.

Answer: D

90. Reference: NFPA 1001, 6.5.3 and 6.5.3(A)(B)

Delmar, *Firefighter's Handbook*, 3rd Edition, 1st Printing, page 361.

IFSTA, *Essentials of Fire Fighting and Fire Department Operations*, 5th Edition, 1st Printing, page 852.

Jones and Bartlett, NFPA, *Fundamentals of Fire Fighter Skills*, 2nd Edition, 1st Printing, page 958.

Answer: A

91. Reference: NFPA 1001, 6.5.3 and 6.5.3(A)(B)

Delmar, *Firefighter's Handbook*, 3rd Edition, 1st Printing, page 367.

IFSTA, *Essentials of Fire Fighting and Fire Department Operations*, 5th Edition, 1st Printing, page 601.

Jones and Bartlett, NFPA, *Fundamentals of Fire Fighter Skills*, 2nd Edition, 1st Printing, page 954.

Answer: B

92. Reference: NFPA 1001, 6.5.3 and 6.5.3(A)(B)

Delmar, *Firefighter's Handbook*, 3rd Edition, 1st Printing, page 357.

IFSTA, *Essentials of Fire Fighting and Fire Department Operations*, 5th Edition, 1st Printing, page 825.

Jones and Bartlett, NFPA, *Fundamentals of Fire Fighter Skills*, 2nd Edition, 1st Printing, page 949.

Answer: C

93. Reference: NFPA 1001, 6.5.3 and 6.5.3(A)(B)

Delmar, *Firefighter's Handbook*, 3rd Edition, 1st Printing, page 363.

IFSTA, *Essentials of Fire Fighting and Fire Department Operations*, 5th Edition, 1st Printing, page 853.

Jones and Bartlett, NFPA, *Fundamentals of Fire Fighter Skills*, 2nd Edition, 1st Printing, page 959.

Answer: C

94. Reference: NFPA 1001, 6.5.3 and 6.5.3(A)(B)

Delmar, *Firefighter's Handbook*, 1st Edition, 1st Printing, pages 363–365.

IFSTA, *Essentials of Fire Fighting and Fire Department Operations*, 5th Edition, 1st Printing, pages 853–854.

Jones and Bartlett, NFPA, *Fundamentals of Fire Fighter Skills*, 2nd Edition, 1st Printing, page 958.

Answer: D

95. Reference: NFPA 1001, 6.5.3 and 6.5.3(A)(B)

Delmar, *Firefighter's Handbook*, 3rd Edition, 1st Printing, page 362.

IFSTA, *Essentials of Fire Fighting and Fire Department Operations*, 5th Edition, 1st Printing, page 856.

Jones and Bartlett, NFPA, *Fundamentals of Fire Fighter Skills*, 2nd Edition, 1st Printing, page 960.

Answer: C

96. Reference: NFPA 1001, 6.5.5 and 6.5.5(A)(B)

Delmar, *Firefighter's Handbook*, 3rd Edition, 1st Printing, page 316.

IFSTA, *Essentials of Fire Fighting and Fire Department Operations*, 5th Edition, 1st Printing, page 713.

Jones and Bartlett, NFPA, *Fundamentals of Fire Fighter Skills*, 2nd Edition, 1st Printing, page 479.

Answer: B

97. Reference: NFPA 1001, 6.5.5 and 6.5.5(A)(B)

Delmar, *Firefighter's Handbook*, 3rd Edition, 1st Printing, page 316.

IFSTA, *Essentials of Fire Fighting and Fire Department Operations*, 5th Edition, 1st Printing, page 713.

Jones and Bartlett, NFPA, *Fundamentals of Fire Fighter Skills*, 2nd Edition, 1st Printing, page 479.

Answer: A

98. Reference: NFPA 1001, 6.5.5 and 6.5.5(A)(B)

Delmar, *Firefighter's Handbook*, 3rd Edition, 1st Printing, page 316.

IFSTA, *Essentials of Fire Fighting and Fire Department Operations*, 5th Edition, 1st Printing, page 713.

Jones and Bartlett, NFPA, *Fundamentals of Fire Fighter Skills*, 2nd Edition, 1st Printing, page 479.

Answer: C

99. Reference: NFPA 1001, 6.5.5 and 6.5.5(A)(B)

Delmar, *Firefighter's Handbook*, 3rd Edition, 1st Printing, page 315.

IFSTA, *Essentials of Fire Fighting and Fire Department Operations*, 5th Edition, 1st Printing, page 680.

Jones and Bartlett, NFPA, *Fundamentals of Fire Fighter Skills*, 2nd Edition, 1st Printing, page 476.

Answer: B

100. Reference: NFPA 1001, 6.5.5 and 6.5.5(A)(B)

Delmar, *Firefighter's Handbook*, 3rd Edition, 1st Printing, page 316.

IFSTA, *Essentials of Fire Fighting and Fire Department Operations*, 5th Edition, 1st Printing, page 682.

Jones and Bartlett, NFPA, *Fundamentals of Fire Fighter Skills*, 2nd Edition, 1st Printing, page 479.

Answer: B

Don't forget to enter the information on your Personal Progress Plotter and answer the Yes and No question at the end of the Examination. This step is extremely important for the successful completion of the Systematic Approach to Examination Preparation!

Examination II-2 Answer Key

Directions

Follow these steps carefully for completing the feedback part of SAEP:

1. After calculating your score, look up the answers for the examination items you missed as well as those on which you guessed, even if you guessed correctly. If you are guessing, it means the answer is not perfectly clear. In this process, we are committed to making you as knowledgeable as possible.

2. Enter the number of missed and guessed examination items in the blanks on your Personal Progress Plotter.

3. Highlight the answer in the reference materials. Read the paragraph preceding and the paragraph following the one in which the correct answer is located. Enter the paragraph number and page number next to the guessed or missed examination item on your examination. Count any part of a paragraph at the beginning of the page as one paragraph until you reach the paragraph containing your highlighted answer. This step will help you locate and review your missed and guessed examination items later in the process. This step is essential to learning the material in context and by association. These learning techniques (context/association) are the very backbone of the SAEP approach.

4. Once you have completed the feedback step, you may proceed to the next examination.

1. Reference: NFPA 1001, 6.1.1 and 6.1.2
   Delmar, *Firefighter's Handbook*, 3rd Edition, 1st Printing, pages 42–44.
   IFSTA, *Essentials of Fire Fighting and Fire Department Operations*, 5th Edition, 1st Printing, page 37.
   Jones and Bartlett, NFPA, *Fundamentals of Fire Fighter Skills*, 2nd Edition, 1st Printing, page 110.
   Answer: A

2. Reference: NFPA 1001, 6.1.1 and 6.1.2
   Delmar, *Firefighter's Handbook*, 3rd Edition, 1st Printing, page 39.
   IFSTA, *Essentials of Fire Fighting and Fire Department Operations*, 5th Edition, 1st Printing, page 70.
   Jones and Bartlett, NFPA, *Fundamentals of Fire Fighter Skills*, 2nd Edition, 1st Printing, page 118.
   Answer: B

3. Reference: NFPA 1001, 6.1.1 and 6.1.2
   Delmar, *Firefighter's Handbook*, 3rd Edition, 1st Printing, page 39.
   IFSTA, *Essentials of Fire Fighting and Fire Department Operations*, 5th Edition, 1st Printing, page 19.
   Jones and Bartlett, NFPA, *Fundamentals of Fire Fighter Skills*, 2nd Edition, 1st Printing, page 107.
   Answer: C

4. Reference: NFPA 1001, 6.1.1 and 6.1.2
Delmar, *Firefighter's Handbook*, 3rd Edition, 1st Printing, page 46.
IFSTA, *Essentials of Fire Fighting and Fire Department Operations*, 5th Edition, 1st Printing, page 36.
Jones and Bartlett, NFPA, *Fundamentals of Fire Fighter Skills*, 2nd Edition, 1st Printing, page 110.
Answer: D

5. Reference: NFPA 1001, 6.1.1 and 6.1.2
Delmar, *Firefighter's Handbook*, 3rd Edition, 1st Printing, page 40.
IFSTA, *Essentials of Fire Fighting and Fire Department Operations*, 5th Edition, 1st Printing, page 946.
Jones and Bartlett, NFPA, *Fundamentals of Fire Fighter Skills*, 2nd Edition, 1st Printing, page 119.
Answer: C

6. References: NFPA 1001, 6.1.1 and 6.1.2
Delmar, *Firefighter's Handbook*, 3rd Edition, 1st Printing, pages 39 and 40.
IFSTA, *Essentials of Fire Fighting and Fire Department Operations*, 5th Edition, 1st Printing, page 792 and 946.
Jones and Bartlett, NFPA, *Fundamentals of Fire Fighter Skills*, 2nd Edition, 1st Printing, pages 119–121.
Answer: D

7. Reference: NFPA 1001, 6.1.1 and 6.1.2
Delmar, *Firefighter's Handbook*, 3rd Edition, 1st Printing, page 42.
IFSTA, *Essentials of Fire Fighting and Fire Department Operations*, 5th Edition, 1st Printing, page 35.
Jones and Bartlett, NFPA, *Fundamentals of Fire Fighter Skills*, 2nd Edition, 1st Printing, page 106.
Answer: D

8. Reference: NFPA 1001, 6.1.1 and 6.1.2
Delmar, *Firefighter's Handbook*, 3rd Edition, 1st Printing, page 46.
IFSTA, *Essentials of Fire Fighting and Fire Department Operations*, 5th Edition, 1st Printing, page 36.
Jones and Bartlett, NFPA, *Fundamentals of Fire Fighter Skills*, 2nd Edition, 1st Printing, page 110.
Answer: C

9. Reference: NFPA 1001, 6.1.1 and 6.1.2
Delmar, *Firefighter's Handbook*, 3rd Edition, 1st Printing, pages 43 and 44.
IFSTA, *Essentials of Fire Fighting and Fire Department Operations*, 5th Edition, 1st Printing, pages 37–38.
Jones and Bartlett, NFPA, *Fundamentals of Fire Fighter Skills*, 2nd Edition, 1st Printing, page 115.
Answer: B

10. Reference: NFPA 1001, 6.1.1 and 6.1.2

Delmar, *Firefighter's Handbook*, 3rd Edition, 1st Printing, page 46.

IFSTA, *Essentials of Fire Fighting and Fire Department Operations*, 5th Edition, 1st Printing, page 39.

Jones and Bartlett, NFPA, *Fundamentals of Fire Fighter Skills*, 2nd Edition, 1st Printing, page 112.

Answer: A

11. Reference: NFPA 1001, 6.1.1, 6.1.2, 6.3.2 and 6.3.2(A)(B)

Delmar, *Firefighter's Handbook*, 3rd Edition, 1st Printing, page 696.

IFSTA, *Essentials of Fire Fighting and Fire Department Operations*, 5th Edition, 1st Printing, pages 96, 570–571.

Jones and Bartlett, NFPA, *Fundamentals of Fire Fighter Skills*, 2nd Edition, 1st Printing, page 113.

Answer: B

12. Reference: NFPA 1001, 6.1.1 and 6.1.2

Delmar, *Firefighter's Handbook*, 3rd Edition, 1st Printing, page 46.

IFSTA, *Essentials of Fire Fighting and Fire Department Operations*, 5th Edition, 1st Printing, page 39.

Jones and Bartlett, NFPA, *Fundamentals of Fire Fighter Skills*, 2nd Edition, 1st Printing, page 114.

Answer: C

13. Reference: NFPA 1001, 6.1.1, 6.1.2, and 6.3.2(A)

Delmar, *Firefighter's Handbook*, 3rd Edition, 1st Printing, page 118.

IFSTA, *Essentials of Fire Fighting and Fire Department Operations*, 5th Edition, 1st Printing, page 58.

Jones and Bartlett, NFPA, *Fundamentals of Fire Fighter Skills*, 2nd Edition, 1st Printing, page 532.

Answer: B

14. Reference: NFPA 1001, 6.1.1, 6.1.2, and 6.3.2(A)

Delmar, *Firefighter's Handbook*, 3rd Edition, 1st Printing, page 130.

IFSTA, *Essentials of Fire Fighting and Fire Department Operations,* 5th Edition, 1st Printing, page 58.

Jones and Bartlett, NFPA, *Fundamentals of Fire Fighter Skills*, 2nd Edition, 1st Printing, page 532.

Answer: C

15. Reference: NFPA 1001, 6.2.1 and 6.2.1(A)(B)

Delmar, *Firefighter's Handbook*, 3rd Edition, 1st Printing, page 78.

IFSTA, *Essentials of Fire Fighting and Fire Department Operations*, 5th Edition, 1st Printing, page 949.

Jones and Bartlett, NFPA, *Fundamentals of Fire Fighter Skills*, 2nd Edition, 1st Printing, page 97.

Answer: D

16. Reference: NFPA 1001, 6.2.2 and 6.2.2(A)(B)

Delmar, *Firefighter's Handbook*, 3rd Edition, 1st Printing, page 76.

IFSTA, *Essentials of Fire Fighting and Fire Department Operations*, 5th Edition, 1st Printing, page 946.

Jones and Bartlett, NFPA, *Fundamentals of Fire Fighter Skills*, 2nd Edition, 1st Printing, page 283.

Answer: B

17. Reference: NFPA 1001, 6.2.2 and 6.2.2(A)(B)

Delmar, *Firefighter's Handbook*, 3rd Edition, 1st Printing, page 836.

IFSTA, *Essentials of Fire Fighting and Fire Department Operations*, 5th Edition, 1st Printing, page 949.

Jones and Bartlett, NFPA, *Fundamentals of Fire Fighter Skills*, 2nd Edition, 1st Printing, page 535.

Answer: D

18. Reference: NFPA 1001, 6.3.1 and 6.3.1(A)(B)

Delmar, *Firefighter's Handbook*, 3rd Edition, 1st Printing, page 338.

IFSTA, *Essentials of Fire Fighting and Fire Department Operations*, 5th Edition, 1st Printing, page 734.

Jones and Bartlett, NFPA, *Fundamentals of Fire Fighter Skills*, 2nd Edition, 1st Printing, page 633.

Answer: C

19. Reference: NFPA 1001, 6.3.1 and 6.3.1(A)(B)

Delmar, *Firefighter's Handbook*, 3rd Edition, 1st Printing, page 339.

IFSTA, *Essentials of Fire Fighting and Fire Department Operations*, 5th Edition, 1st Printing, page 738.

Jones and Bartlett, NFPA, *Fundamentals of Fire Fighter Skills*, 2nd Edition, 1st Printing, page 522.

Answer: C

20. Reference: NFPA 1001, 6.3.1 and 6.3.1(A)(B)

Delmar, *Firefighter's Handbook*, 3rd Edition, 1st Printing, page 341.

IFSTA, *Essentials of Fire Fighting and Fire Department Operations*, 5th Edition, 1st Printing, page 1323.

Jones and Bartlett, NFPA, *Fundamentals of Fire Fighter Skills*, 2nd Edition, 1st Printing, page 522.

Answer: A

21. Reference: NFPA 1001, 6.3.1 and 6.3.1(A)(B)

Delmar, *Firefighter's Handbook*, 3rd Edition, 1st Printing, page 105.

IFSTA, *Essentials of Fire Fighting and Fire Department Operations*, 5th Edition, 1st Printing, page 774.

Jones and Bartlett, NFPA, *Fundamentals of Fire Fighter Skills*, 2nd Edition, 1st Printing, page 633.

Answer: A

22. References: NFPA 1001, 6.3.1 and 6.3.1(A)(B)

Delmar, *Firefighter's Handbook*, 2nd Edition, 1st Printing, page 344.

IFSTA, *Essentials of Fire Fighting and Fire Department Operations*, 5th Edition, 1st Printing, pages 737 and 745.

Jones and Bartlett, NFPA, *Fundamentals of Fire Fighter Skills*, 2nd Edition, 1st Printing, page 523.

Answer: D

23. References: NFPA 1001, 6.3.1 and 6.3.1(A)

Delmar, *Firefighter's Handbook*, 3rd Edition, 1st Printing, page 338.

IFSTA, *Essentials of Fire Fighting and Fire Department Operations*, 5th Edition, 1st Printing, page 734.

Jones and Bartlett, NFPA, *Fundamentals of Fire Fighter Skills*, 2nd Edition, 1st Printing, page 521.

Answer: D

24. Reference: NFPA 1001, 6.3.2 and 6.3.2(A)(B)

Delmar, *Firefighter's Handbook*, 3rd Edition, 1st Printing, page 708.

IFSTA, *Essentials of Fire Fighting and Fire Department Operations*, 5th Edition, 1st Printing, page 783.

Jones and Bartlett, NFPA, *Fundamentals of Fire Fighter Skills*, 2nd Edition, 1st Printing, pages 636–637.

Answer: D

25. Reference: NFPA 1001, 6.3.2 and 6.3.2(A)(B)

Delmar, *Firefighter's Handbook*, 3rd Edition, 1st Printing, page 406.

IFSTA, *Essentials of Fire Fighting and Fire Department Operations*, 5th Edition, 1st Printing, page 557.

Jones and Bartlett, NFPA, *Fundamentals of Fire Fighter Skills*, 2nd Edition, 1st Printing, page 165.

Answer: C

26. Reference: NFPA 1001, 6.3.2 and 6.3.2(A)(B)

Delmar, *Firefighter's Handbook*, 3rd Edition, 1st Printing, pages 744–745.

IFSTA, *Essentials of Fire Fighting and Fire Department Operations*, 5th Edition, 1st Printing, page 884.

Jones and Bartlett, NFPA, *Fundamentals of Fire Fighter Skills*, 2nd Edition, 1st Printing, page 576.

Answer: D

27. Reference: NFPA 1001, 6.3.2 and 6.3.2(A)(B)

Delmar, *Firefighter's Handbook*, 3rd Edition, 1st Printing, pages 90–91.

IFSTA, *Essentials of Fire Fighting and Fire Department Operations*, 5th Edition, 1st Printing, page 93.

Jones and Bartlett, NFPA, *Fundamentals of Fire Fighter Skills*, 2nd Edition, 1st Printing, page 129.

Answer: C

28. Reference: NFPA 1001, 6.3.2 and 6.3.2(A)(B)

Delmar, *Firefighter's Handbook*, 3rd Edition, 1st Printing, page 655.

IFSTA, *Essentials of Fire Fighting and Fire Department Operations*, 5th Edition, 1st Printing, page 577.

Jones and Bartlett, NFPA, *Fundamentals of Fire Fighter Skills*, 2nd Edition, 1st Printing, pages 416–417.

Answer: A

29. Reference: NFPA 1001, 6.3.2 and 6.3.2(A)(B)

Delmar, *Firefighter's Handbook*, 3rd Edition, 1st Printing, page 326.

IFSTA, *Essentials of Fire Fighting and Fire Department Operations*, 5th Edition, 1st Printing, page 727.

Jones and Bartlett, NFPA, *Fundamentals of Fire Fighter Skills*, 2nd Edition, 1st Printing, page 517.

Answer: B

30. Reference: NFPA 1001, 6.3.2 and 6.3.2(A)(B)

Delmar, *Firefighter's Handbook*, 3rd Edition, 1st Printing, page 404.

IFSTA, *Essentials of Fire Fighting and Fire Department Operations*, 5th Edition, 1st Printing, page 156.

Jones and Bartlett, NFPA, *Fundamentals of Fire Fighter Skills*, 2nd Edition, 1st Printing, page 423.

Answer: A

31. Reference: NFPA 1001, 6.3.2 and 6.3.2(A)(B)

Delmar, *Firefighter's Handbook*, 3rd Edition, 1st Printing, page 398.

IFSTA, *Essentials of Fire Fighting and Fire Department Operations*, 5th Edition, 1st Printing, page 567.

Jones and Bartlett, NFPA, *Fundamentals of Fire Fighter Skills*, 2nd Edition, 1st Printing, page 162.

Answer: C

32. Reference: NFPA 1001, 6.3.2 and 6.3.2(A)(B)

Delmar, *Firefighter's Handbook*, 3rd Edition, 1st Printing, pages 103 and 163.

IFSTA, *Essentials of Fire Fighting and Fire Department Operations*, 5th Edition, 1st Printing, page 108.

Jones and Bartlett, NFPA, *Fundamentals of Fire Fighter Skills*, 2nd Edition, 1st Printing, page 131.

Answer: D

33. Reference: NFPA 1001, 6.3.2 and 6.3.2(A)(B)

Delmar, *Firefighter's Handbook*, 3rd Edition, 1st Printing, page 524.

IFSTA, *Essentials of Fire Fighting and Fire Department Operations*, 5th Edition, 1st Printing, page 309.

Jones and Bartlett, NFPA, *Fundamentals of Fire Fighter Skills*, 2nd Edition, 1st Printing, page 369.

Answer: A

34. Reference: NFPA 1001, 6.3.2 and 6.3.2(A)(B)

Delmar, *Firefighter's Handbook*, 3rd Edition, 1st Printing, page 660.

IFSTA, *Essentials of Fire Fighting and Fire Department Operations*, 5th Edition, 1st Printing, page 567.

Jones and Bartlett, NFPA, *Fundamentals of Fire Fighter Skills*, 2nd Edition, 1st Printing, page 432.

Answer: A

35. Reference: NFPA 1001, 6.3.2 and 6.3.2(A)(B)

Delmar, *Firefighter's Handbook*, 3rd Edition, 1st Printing, pages 42, 44, and 532.

IFSTA, *Essentials of Fire Fighting and Fire Department Operations*, 5th Edition, 1st Printing, page 323.

Jones and Bartlett, NFPA, *Fundamentals of Fire Fighter Skills*, 2nd Edition, 1st Printing, page 536.

Answer: C

36. Reference: NFPA 1001, 6.3.2 and 6.3.2(A)(B)

Delmar, *Firefighter's Handbook*, 3rd Edition, 1st Printing, pages 141 and 151.

IFSTA, *Essentials of Fire Fighting and Fire Department Operations*, 5th Edition, 1st Printing, page 881.

Jones and Bartlett, NFPA, *Fundamentals of Fire Fighter Skills*, 2nd Edition, 1st Printing, page 573.

Answer: C

37. References: NFPA 1001, 6.3.2 and 6.3.2(A)(B)

Delmar, *Firefighter's Handbook*, 3rd Edition, 1st Printing, page 391.

IFSTA, *Essentials of Fire Fighting and Fire Department Operations*, 5th Edition, 1st Printing, page 142.

Jones and Bartlett, NFPA, *Fundamentals of Fire Fighter Skills*, 2nd Edition, 1st Printing, pages 153–154.

Answer: D

38. References: NFPA 1001, 6.3.2 and 6.3.2(A)(B)

Delmar, *Firefighter's Handbook*, 3rd Edition, 1st Printing, pages 390–391.

IFSTA, *Essentials of Fire Fighting and Fire Department Operations*, 5th Edition, 1st Printing, page 146.

Jones and Bartlett, NFPA, *Fundamentals of Fire Fighter Skills*, 2nd Edition, 1st Printing, page 154.

Answer: C

39. References: NFPA 1001, 6.3.2 and 6.3.2(A)(B)

Delmar, *Firefighter's Handbook*, 3rd Edition, 1st Printing, page 391.

IFSTA, *Essentials of Fire Fighting and Fire Department Operations*, 5th Edition, 1st Printing, page 144.

Jones and Bartlett, NFPA, *Fundamentals of Fire Fighter Skills*, 2nd Edition, 1st Printing, page 154.

Answer: C

40. Reference: NFPA 1001, 6.3.2 and 6.3.2(A)(B)

Delmar, *Firefighter's Handbook*, 3rd Edition, 1st Printing, page 395.

IFSTA, *Essentials of Fire Fighting and Fire Department Operations*, 5th Edition, page 147.

Jones and Bartlett, NFPA, *Fundamentals of Fire Fighter Skills*, 2nd Edition, 1st Printing, page 158.

Answer: B

41. Reference: NFPA 1001, 6.3.3 and 6.3.3(A)

Delmar, *Firefighter's Handbook*, 3rd Edition, 1st Printing, pages 105–106 and 890.

IFSTA, *Essentials of Fire Fighting and Fire Department Operations*, 5th Edition, 1st Printing, page 774.

Jones and Bartlett, NFPA, *Fundamentals of Fire Fighter Skills*, 2nd Edition, 1st Printing, page 142.

Answer: C

42. Reference: NFPA 1001, 6.3.3 and 6.3.3(A)(B)

Delmar, *Firefighter's Handbook*, 3rd Edition, 1st Printing, page 1001.

IFSTA, *Essentials of Fire Fighting and Fire Department Operations*, 5th Edition, 1st Printing, page 774.

Jones and Bartlett, NFPA, *Fundamentals of Fire Fighter Skills*, 2nd Edition, 1st Printing, page 635.

Answer: B

43. Reference: NFPA 1001, 6.3.3 and 6.3.3(A)(B)

Delmar, *Firefighter's Handbook*, 3rd Edition, 1st Printing, page 893.

IFSTA, *Essentials of Fire Fighting and Fire Department Operations*, 5th Edition, 1st Printing, page 776.

Jones and Bartlett, NFPA, *Fundamentals of Fire Fighter Skills*, 2nd Edition, 1st Printing, page 635.

Answer: A

44. References: NFPA 1001, 6.3.3 and 6.3.3(A)

Delmar, *Firefighter's Handbook*, 3rd Edition, 1st Printing, page 908.

IFSTA, *Essentials of Fire Fighting and Fire Department Operations*, 5th Edition, 1st Printing, page 779.

Jones and Bartlett, NFPA, *Fundamentals of Fire Fighter Skills*, 2nd Edition, 1st Printing, page 141.

Answer: B

45. Reference: NFPA 1001, 6.3.3 and 6.3.3(A)(B)

Delmar, *Firefighter's Handbook*, 3rd Edition, 1st Printing, page 908.

IFSTA, *Essentials of Fire Fighting and Fire Department Operations*, 5th Edition, 1st Printing, page 779.

Jones and Bartlett, NFPA, *Fundamentals of Fire Fighter Skills*, 2nd Edition, 1st Printing, page 141.

Answer: C

46. Reference: NFPA 1001, 6.3.4 and 6.3.4(A)(B)

Delmar, *Firefighter's Handbook*, 3rd Edition, 1st Printing, page 746.

IFSTA, *Essentials of Fire Fighting and Fire Department Operations*, 5th Edition, 1st Printing, pages 911–913.

Jones and Bartlett, NFPA, *Fundamentals of Fire Fighter Skills*, 2nd Edition, 1st Printing, pages 981–983.

Answer: D

47. Reference: NFPA 1001, 6.3.4 and 6.3.4(A)(B)

Delmar, *Firefighter's Handbook*, 3rd Edition, 1st Printing, page 747.

IFSTA, *Essentials of Fire Fighting and Fire Department Operations*, 5th Edition, 1st Printing, page 910.

Jones and Bartlett, NFPA, *Fundamentals of Fire Fighter Skills*, 2nd Edition, 1st Printing, page 981.

Answer: B

48. Reference: NFPA 1001, 6.3.4 and 6.3.4(A)(B)

Delmar, *Firefighter's Handbook*, 3rd Edition, 1st Printing, page 743.

IFSTA, *Essentials of Fire Fighting and Fire Department Operations*, 5th Edition, 1st Printing, page 918.

Jones and Bartlett, NFPA, *Fundamentals of Fire Fighter Skills*, 2nd Edition, 1st Printing, page 575.

Answer: D

49. Reference: NFPA 1001, 6.4.1 and 6.4.1(A)(B)

Delmar, *Firefighter's Handbook*, 3rd Edition, 1st Printing, pages 555–556.

IFSTA, *Essentials of Fire Fighting and Fire Department Operations*, 5th Edition, 1st Printing, page 358.

Jones and Bartlett, NFPA, *Fundamentals of Fire Fighter Skills*, 2nd Edition, 1st Printing, page 735.

Answer: B

50. Reference: NFPA 1001, 6.4.1 and 6.4.1(A)(B)

Delmar, *Firefighter's Handbook*, 3rd Edition, 1st Printing, page 553.

IFSTA, *Essentials of Fire Fighting and Fire Department Operations*, 5th Edition, 1st Printing, page 353.

Jones and Bartlett, NFPA, *Fundamentals of Fire Fighter Skills*, 2nd Edition, 1st Printing, page 733.

Answer: B

51. Reference: NFPA 1001, 6.4.1 and 6.4.1(A)(B)

Delmar, *Firefighter's Handbook*, 3rd Edition, 1st Printing, page 551.

IFSTA, *Essentials of Fire Fighting and Fire Department Operations*, 5th Edition, 1st Printing, page 347.

Jones and Bartlett, NFPA, *Fundamentals of Fire Fighter Skills*, 2nd Edition, 1st Printing, page 726.

Answer: D

52. Reference: NFPA 1001, 6.4.1 and 6.4.1(A)(B)
Delmar, *Firefighter's Handbook*, 3rd Edition, 1st Printing, page 545.
IFSTA, *Essentials of Fire Fighting and Fire Department Operations*, 5th Edition, 1st Printing, page 347.
Jones and Bartlett, NFPA, *Fundamentals of Fire Fighter Skills*, 2nd Edition, 1st Printing, page 726.
Answer: B

53. Reference: NFPA 1001, 6.4.1 and 6.4.1(A)(B)
Delmar, *Firefighter's Handbook*, 3rd Edition, 1st Printing, page 554.
IFSTA, *Essentials of Fire Fighting and Fire Department Operations*, 5th Edition, 1st Printing, page 355.
Jones and Bartlett, NFPA, *Fundamentals of Fire Fighter Skills*, 2nd Edition, 1st Printing, page 737.
Answer: C

54. Reference: NFPA 1001, 6.4.1 and 6.4.1(A)(B)
Delmar, *Firefighter's Handbook*, 3rd Edition, 1st Printing, page 557.
IFSTA, *Essentials of Fire Fighting and Fire Department Operations,* 5th Edition, 1st Printing, pages 355–356.
Jones and Bartlett, NFPA, *Fundamentals of Fire Fighter Skills*, 2nd Edition, 1st Printing, page 743.
Answer: B

55. Reference: NFPA 1001, 6.4.1, 6.4.1(A)(B), 6.4.2, and 6.4.2(A)(B)
Delmar, *Firefighter's Handbook*, 3rd Edition, 1st Printing, page 58.
IFSTA, *Essentials of Fire Fighting and Fire Department Operations,* 5th Edition, 1st Printing, page 335.
Answer: B

56. Reference: NFPA 1001, 6.4.1 and 6.4.1(A)(B)
Delmar, *Firefighter's Handbook*, 3rd Edition, 1st Printing, page 555.
IFSTA, *Essentials of Fire Fighting and Fire Department Operations*, 5th Edition, 1st Printing, page 356.
Jones and Bartlett, NFPA, *Fundamentals of Fire Fighter Skills*, 2nd Edition, 1st Printing, page 733.
Answer: D

57. References: NFPA 1001, 6.4.1, and 6.4.1(A)(B)
Delmar, *Firefighter's Handbook*, 3rd Edition, 1st Printing, page 553.
IFSTA, *Essentials of Fire Fighting and Fire Department Operations,* 5th Edition, 1st Printing, page 353.
Jones and Bartlett, NFPA, *Fundamentals of Fire Fighter Skills*, 2nd Edition, 1st Printing, page 729.
Answer: D

58. References: NFPA 1001, 6.4.1 and 6.4.1(A)(B)
Delmar, *Firefighter's Handbook*, 3rd Edition, 1st Printing, page 550.
IFSTA, *Essentials of Fire Fighting and Fire Department Operations*, 5th Edition, 1st Printing, page 402.
Jones and Bartlett, NFPA, *Fundamentals of Fire Fighter Skills*, 2nd Edition, 1st Printing, page 226.
Answer: B

59. Reference: NFPA 1001, 6.4.1 and 6.4.1(A)(B)

Delmar, *Firefighter's Handbook*, 3rd Edition, 1st Printing, page 554.

IFSTA, *Essentials of Fire Fighting and Fire Department Operations*, 5th Edition, pages 354 and 355.

Jones and Bartlett, NFPA, *Fundamentals of Fire Fighter Skills*, 2nd Edition, 1st Printing, page 740.

Answer: D

60. Reference: NFPA 1001, 6.4.2 and 6.4.2(A)(B)

Delmar, *Firefighter's Handbook*, 3rd Edition, 1st Printing, page 570.

IFSTA, *Essentials of Fire Fighting and Fire Department Operations*, 5th Edition, 1st Printing, page 365.

Jones and Bartlett, NFPA, *Fundamentals of Fire Fighter Skills*, 2nd Edition, 1st Printing, page 763.

Answer: A

61. Reference: NFPA 1001, 6.4.2 and 6.4.2(A)(B)

Delmar, *Firefighter's Handbook*, 3rd Edition, 1st Printing, page 573.

IFSTA, *Essentials of Fire Fighting and Fire Department Operations*, 5th Edition, 1st Printing, pages 367 and 800.

Jones and Bartlett, NFPA, *Fundamentals of Fire Fighter Skills*, 2nd Edition, 1st Printing, page 759.

Answer: A

62. Reference: NFPA 1001, 6.4.2 and 6.4.2(A)(B)

Delmar, *Firefighter's Handbook*, 3rd Edition, 1st Printing, page 533.

IFSTA, *Essentials of Fire Fighting and Fire Department Operations*, 5th Edition, 1st Printing, page 328.

Jones and Bartlett, NFPA, *Fundamentals of Fire Fighter Skills*, 2nd Edition, 1st Printing, page 33.

Answer: B

63. Reference: NFPA 1001, 6.4.2 and 6.4.2(A)(B)

Delmar, *Firefighter's Handbook*, 3rd Edition, 1st Printing, page 557.

IFSTA, *Essentials of Fire Fighting and Fire Department Operations*, 5th Edition, 1st Printing, page 74.

Jones and Bartlett, NFPA, *Fundamentals of Fire Fighter Skills*, 2nd Edition, page 753.

Answer: B

64. Reference: NFPA 1001, 6.5.1 and 6.5.1(A)(B)

Delmar, *Firefighter's Handbook*, 3rd Edition, 1st Printing, page 391.

IFSTA, *Essentials of Fire Fighting and Fire Department Operations,* 5th Edition, 1st Printing, page 144.

Jones and Bartlett, NFPA, *Fundamentals of Fire Fighter Skills*, 2nd Edition, 1st Printing, page 154.

Answer: B

65. Reference: NFPA 1001, 6.5.1 and 6.5.1(A)(B)

Delmar, *Firefighter's Handbook*, 3rd Edition, 1st Printing, pages 763–764.

IFSTA, *Essentials of Fire Fighting and Fire Department Operations*, 5th Edition, 1st Printing, pages 971–972.

Jones and Bartlett, NFPA, *Fundamentals of Fire Fighter Skills*, 2nd Edition, 1st Printing, pages 650–651.

Answer: B

66. Reference: NFPA 1001, 6.5.1 and 6.5.1(A)

Delmar, *Firefighter's Handbook*, 3rd Edition, 1st Printing, page 404.

IFSTA, *Essentials of Fire Fighting and Fire Department Operations*, 5th Edition, 1st Printing, page 156.

Jones and Bartlett, NFPA, *Fundamentals of Fire Fighter Skills*, 2nd Edition, 1st Printing, page 423.

Answer: B

67. References: NFPA 1001, 6.5.1 and 6.5.1(A)(B)

Delmar, *Firefighter's Handbook*, 3rd Edition, 1st Printing, page 395.

IFSTA, *Essentials of Fire Fighting and Fire Department Operations*, 5th Edition, 1st Printing, pages 147.

Jones and Bartlett, NFPA, *Fundamentals of Fire Fighter Skills*, 2nd Edition, 1st Printing, page 648.

Answer: B

68. Reference: NFPA 1001, 6.5.1 and 6.5.1(A)(B)

Delmar, *Firefighter's Handbook*, 3rd Edition, 1st Printing, page 756.

IFSTA, *Essentials of Fire Fighting and Fire Department Operations*, 5th Edition, 1st Printing, page 961.

Jones and Bartlett, NFPA, *Fundamentals of Fire Fighter Skills*, 2nd Edition, 1st Printing, pages 643–644.

Answer: C

69. Reference: NFPA 1001, 6.5.1 and 6.5.1(A)(B)

Delmar, *Firefighter's Handbook*, 3rd Edition, 1st Printing, pages 775–776.

IFSTA, *Essentials of Fire Fighting and Fire Department Operations*, 5th Edition, 1st Printing, page 980.

Jones and Bartlett, NFPA, *Fundamentals of Fire Fighter Skills*, 2nd Edition, 1st Printing, page 926.

Answer: D

70. Reference: NFPA 1001, 6.5.1 and 6.5.1(A)(B)

Delmar, *Firefighter's Handbook*, 3rd Edition, 1st Printing, page 782.

IFSTA, *Essentials of Fire Fighting and Fire Department Operations*, 5th Edition, 1st Printing, page 972.

Jones and Bartlett, NFPA, *Fundamentals of Fire Fighter Skills*, 2nd Edition, 1st Printing, page 645.

Answer: D

71. Reference: NFPA 1001, 6.5.1 and 6.5.1(A)(B)

Delmar, *Firefighter's Handbook*, 3rd Edition, 1st Printing, page 782.

IFSTA, *Essentials of Fire Fighting and Fire Department Operations*, 5th Edition, 1st Printing, page 961.

Jones and Bartlett, NFPA, *Fundamentals of Fire Fighter Skills*, 2nd Edition, 1st Printing, page 658.

Answer: D

72. Reference: NFPA 1001, 6.5.1 and 6.5.1(A)(B)

Delmar, *Firefighter's Handbook*, 3rd Edition, 1st Printing, page 762.

IFSTA, *Essentials of Fire Fighting and Fire Department Operations*, 5th Edition, 1st Printing, page 250.

Jones and Bartlett, NFPA, *Fundamentals of Fire Fighter Skills*, 2nd Edition, 1st Printing, page 207.

Answer: B

73. Reference: NFPA 1001, 6.5.1 and 6.5.1(A)(B)

Delmar, *Firefighter's Handbook*, 2nd Edition, 1st Printing, page 775.

IFSTA, *Essentials of Fire Fighting and Fire Department Operations*, 5th Edition, 1st Printing, page 976.

Jones and Bartlett, NFPA, *Fundamentals of Fire Fighter Skills*, 2nd Edition, 1st Printing, page 923.

Answer: D

74. Reference: NFPA 1001, 6.5.2 and 6.5.2(A)(B)

Delmar, *Firefighter's Handbook*, 3rd Edition, 1st Printing, page 777.

IFSTA, *Essentials of Fire Fighting and Fire Department Operations*, 5th Edition, 1st Printing, page 985.

Jones and Bartlett, NFPA, *Fundamentals of Fire Fighter Skills*, 2nd Edition, 1st Printing, page 929.

Answer: D

75. Reference: NFPA 1001, 6.5.3 and 6.5.3(A)(B)

Delmar, *Firefighter's Handbook*, 3rd Edition, 1st Printing, page 353.

IFSTA, *Essentials of Fire Fighting and Fire Department Operations*, 5th Edition, 1st Printing, page 827.

Jones and Bartlett, NFPA, *Fundamentals of Fire Fighter Skills*, 2nd Edition, 1st Printing, page 943.

Answer: B

76. Reference: NFPA 1001, 6.5.3 and 6.5.3(A)(B)

Delmar, *Firefighter's Handbook*, 3rd Edition, 1st Printing, page 353.

IFSTA, *Essentials of Fire Fighting and Fire Department Operations*, 5th Edition, 1st Printing, pages 832–833.

Jones and Bartlett, NFPA, *Fundamentals of Fire Fighter Skills*, 2nd Edition, 1st Printing, page 940.

Answer: C

77. Reference: NFPA 1001, 6.5.3 and 6.5.3(A)(B)

Delmar, *Firefighter's Handbook*, 3rd Edition, 1st Printing, page 364.

IFSTA, *Essentials of Fire Fighting and Fire Department Operations*, 5th Edition, 1st Printing, page 854.

Jones and Bartlett, NFPA, *Fundamentals of Fire Fighter Skills*, 2nd Edition, 1st Printing, page 959.

Answer: C

78. Reference: NFPA 1001, 6.5.3 and 6.5.3(A)(B)

Delmar, *Firefighter's Handbook*, 3rd Edition, 1st Printing, page 367.

IFSTA, *Essentials of Fire Fighting and Fire Department Operations*, 5th Edition, 1st Printing, page 848.

Jones and Bartlett, NFPA, *Fundamentals of Fire Fighter Skills*, 2nd Edition, 1st Printing, page 954.

Answer: D

79. Reference: NFPA 1001, 6.5.3 and 6.5.3(A)(B)

Delmar, *Firefighter's Handbook*, 3rd Edition, 1st Printing, page 367.

IFSTA, *Essentials of Fire Fighting and Fire Department Operations*, 5th Edition, 1st Printing, page 848.

Jones and Bartlett, NFPA, *Fundamentals of Fire Fighter Skills*, 2nd Edition, 1st Printing, page 954.

Answer: B

80. Reference: NFPA 1001, 6.5.3 and 6.5.3(A)(B)

Delmar, *Firefighter's Handbook*, 3rd Edition, 1st Printing, pages 359–360.

IFSTA, *Essentials of Fire Fighting and Fire Department Operations*, 5th Edition, 1st Printing, page 845.

Jones and Bartlett, NFPA, *Fundamentals of Fire Fighter Skills*, 2nd Edition, 1st Printing, page 950.

Answer: A

81. Reference: NFPA 1001, 6.5.3 and 6.5.3(A)(B)

Delmar, *Firefighter's Handbook*, 3rd Edition, 1st Printing, page 782.

IFSTA, *Essentials of Fire Fighting and Fire Department Operations*, 5th Edition, 1st Printing, page 970.

Jones and Bartlett, NFPA, *Fundamentals of Fire Fighter Skills*, 2nd Edition, 1st Printing, page 645.

Answer: D

82. Reference: NFPA 1001, 6.5.3 and 6.5.3(A)(B)

Delmar, *Firefighter's Handbook*, 3rd Edition, 1st Printing, page 782.

IFSTA, *Essentials of Fire Fighting and Fire Department Operations*, 5th Edition, 1st Printing, page 970.

Jones and Bartlett, NFPA, *Fundamentals of Fire Fighter Skills*, 2nd Edition, 1st Printing, page 643.

Answer: D

83. Reference: NFPA 1001, 6.5.3 and 6.5.3(A)(B)

Delmar, *Firefighter's Handbook*, 3rd Edition, 1st Printing, page 215.

IFSTA, *Essentials of Fire Fighting and Fire Department Operations*, 5th Edition, 1st Printing, page 964.

Jones and Bartlett, NFPA, *Fundamentals of Fire Fighter Skills*, 2nd Edition, 1st Printing, page 658.

Answer: A

84. Reference: NFPA 1001, 6.5.3 and 6.5.3(A)(B)

Delmar, *Firefighter's Handbook*, 3rd Edition, 1st Printing, page 757.

IFSTA, *Essentials of Fire Fighting and Fire Department Operations*, 5th Edition, 1st Printing, page 968.

Jones and Bartlett, NFPA, *Fundamentals of Fire Fighter Skills*, 2nd Edition, 1st Printing, page 645.

Answer: B

85. Reference: NFPA 1001, 6.5.3 and 6.5.3(A)(B)

Delmar, *Firefighter's Handbook*, 3rd Edition, 1st Printing, page 366.

IFSTA, *Essentials of Fire Fighting and Fire Department Operations*, 5th Edition, 1st Printing, page 851.

Jones and Bartlett, NFPA, *Fundamentals of Fire Fighter Skills*, 2nd Edition, 1st Printing, page 956.

Answer: B

86. Reference: NFPA 1001, 6.5.3 and 6.5.3(A)(B)

Delmar, *Firefighter's Handbook*, 3rd Edition, 1st Printing, page 361.

IFSTA, *Essentials of Fire Fighting and Fire Department Operations*, 5th Edition, 1st Printing, page 845.

Jones and Bartlett, NFPA, *Fundamentals of Fire Fighter Skills*, 2nd Edition, 1st Printing, page 954.

Answer: D

87. Reference: NFPA 1001, 6.5.3 and 6.5.3(A)(B)

Delmar, *Firefighter's Handbook*, 2nd Edition, 1st Printing, pages 353, 355, and 357.

IFSTA, *Essentials of Fire Fighting and Fire Department Operations*, 5th Edition, 1st Printing, pages 826–834.

Jones and Bartlett, NFPA, *Fundamentals of Fire Fighter Skills*, 2nd Edition, 1st Printing, pages 942–944.

Answer: A

88. Reference: NFPA 1001, 6.5.3 and 6.5.3(A)(B)

Delmar, *Firefighter's Handbook*, 3rd Edition, 1st Printing, page 353.

IFSTA, *Essentials of Fire Fighting and Fire Department Operations*, 5th Edition, 1st Printing, page 833.

Jones and Bartlett, NFPA, *Fundamentals of Fire Fighter Skills*, 2nd Edition, 1st Printing, page 939.

Answer: D

89. Reference: NFPA 1001, 6.5.3 and 6.5.3(A)(B)
Delmar, *Firefighter's Handbook*, 3rd Edition, 1st Printing, page 64.
IFSTA, *Essentials of Fire Fighting and Fire Department Operations*, 5th Edition, 1st Printing, page 839.
Jones and Bartlett, NFPA, *Fundamentals of Fire Fighter Skills*, 2nd Edition, 1st Printing, page 949.
Answer: D

90. Reference: NFPA 1001, 6.5.3 and 6.5.3(A)(B)
Delmar, *Firefighter's Handbook*, 3rd Edition, 1st Printing, page 64.
IFSTA, *Essentials of Fire Fighting and Fire Department Operations*, 5th Edition, 1st Printing, page 826.
Jones and Bartlett, NFPA, *Fundamentals of Fire Fighter Skills*, 2nd Edition, 1st Printing, page 948.
Answer: C

91. Reference: NFPA 1001, 6.5.3 and 6.5.3(A)(B)
Delmar, *Firefighter's Handbook*, 3rd Edition, 1st Printing, page 355.
IFSTA, *Essentials of Fire Fighting and Fire Department Operations*, 5th Edition, 1st Printing, page 832.
Jones and Bartlett, NFPA, *Fundamentals of Fire Fighter Skills*, 2nd Edition, 1st Printing, page 939.
Answer: A

92. Reference: NFPA 1001, 6.5.3 and 6.5.33(A)(B)
Delmar, *Firefighter's Handbook*, 3rd Edition, 1st Printing, page 353.
IFSTA, *Essentials of Fire Fighting and Fire Department Operations*, 5th Edition, 1st Printing, page 829.
Jones and Bartlett, NFPA, *Fundamentals of Fire Fighter Skills*, 2nd Edition, 1st Printing, page 943.
Answer: D

93. Reference: NFPA 1001, 6.5.3 and 6.5.3(A)
Delmar, *Firefighter's Handbook*, 3rd Edition, 1st Printing, page 64.
IFSTA, *Essentials of Fire Fighting and Fire Department Operations*, 5th Edition, 1st Printing, page 839.
Jones and Bartlett, NFPA, *Fundamentals of Fire Fighter Skills*, 2nd Edition, 1st Printing, page 949.
Answer: D

94. Reference: NFPA 1001, 6.5.3 and 6.5.3(A)
Delmar, *Firefighter's Handbook*, 3rd Edition, 1st Printing, page 64.
IFSTA, *Essentials of Fire Fighting and Fire Department Operations*, 5th Edition, 1st Printing, pages 826, 837, and 839.
Jones and Bartlett, NFPA, *Fundamentals of Fire Fighter Skills*, 2nd Edition, 1st Printing, pages 948–949.
Answer: D

95. Reference: NFPA 1001, 6.5.3 and 6.5.3(A)(B)

Delmar, *Firefighter's Handbook*, 3rd Edition, 1st Printing, page 353.

IFSTA, *Essentials of Fire Fighting and Fire Department Operations*, 5th Edition, 1st Printing, page 828.

Jones and Bartlett, NFPA, *Fundamentals of Fire Fighter Skills*, 2nd Edition, 1st Printing, page 943.

Answer: D

96. Reference: NFPA 1001, 6.5.3 and 6.5.3(A)(B)

Delmar, *Firefighter's Handbook*, 3rd Edition, 1st Printing, page 96.

IFSTA, *Essentials of Fire Fighting and Fire Department Operations*, 5th Edition, 1st Printing, page 112.

Jones and Bartlett, NFPA, *Fundamentals of Fire Fighter Skills*, 2nd Edition, 1st Printing, page 134.

Answer: C

97. Reference: NFPA 1001, 6.5.3 and 6.5.3(A)(B)

Delmar, *Firefighter's Handbook*, 3rd Edition, 1st Printing, page 360.

IFSTA, *Essentials of Fire Fighting and Fire Department Operations*, 5th Edition, 1st Printing, pages 845–846.

Jones and Bartlett, NFPA, *Fundamentals of Fire Fighter Skills*, 2nd Edition, 1st Printing, pages 950–951.

Answer: B

98. Reference: NFPA 1001, 6.5.5 and 6.5.5(A)(B)

Delmar, *Firefighter's Handbook*, 3rd Edition, 1st Printing, page 316.

IFSTA, *Essentials of Fire Fighting and Fire Department Operations*, 5th Edition, 1st Printing, page 713.

Jones and Bartlett, NFPA, *Fundamentals of Fire Fighter Skills*, 2nd Edition, 1st Printing, page 479.

Answer: B

99. Reference: NFPA 1001, 6.5.5 and 6.5.5(A)(B)

Delmar, *Firefighter's Handbook*, 3rd Edition, 1st Printing, page 316.

IFSTA, *Essentials of Fire Fighting and Fire Department Operations*, 5th Edition, 1st Printing, page 713.

Jones and Bartlett, NFPA, *Fundamentals of Fire Fighter Skills*, 2nd Edition, 1st Printing, page 479.

Answer: D

100. Reference: NFPA 1001, 6.5.5 and 6.5.5(A)(B)

Delmar, *Firefighter's Handbook*, 3rd Edition, 1st Printing, page 316.

IFSTA, *Essentials of Fire Fighting and Fire Department Operations*, 5th Edition, 1st Printing, page 680.

Jones and Bartlett, NFPA, *Fundamentals of Fire Fighter Skills*, 2nd Edition, 1st Printing, page 479.

Answer: C

Don't forget to enter the information on your Personal Progress Plotter and answer the Yes and No question at the end of the Examination. This step is extremely important for the successful completion of the Systematic Approach to Examination Preparation!

Examination II-3 Answer Key

Directions

Follow these steps carefully for completing the feedback part of SAEP:

1. After calculating your score, look up the answers for the examination items you missed as well as those on which you guessed, even if you guessed correctly. If you are guessing, it means the answer is not perfectly clear. In this process, we are committed to making you as knowledgeable as possible.

2. Enter the number of missed and guessed examination items in the blanks on your Personal Progress Plotter.

3. Highlight the answer in the reference materials. Read the paragraph preceding and the paragraph following the one in which the correct answer is located. Enter the paragraph number and page number next to the guessed or missed examination item on your examination. Count any part of a paragraph at the beginning of the page as one paragraph until you reach the paragraph containing your highlighted answer. This step will help you locate and review your missed and guessed examination items later in the process. This step is essential to learning the material in context and by association. These learning techniques (context/association) are the very backbone of the SAEP approach.

4. Congratulations! You have completed the examination and feedback parts of SAEP when you have highlighted your guessed and missed examination items for this examination.

Proceed to Phases III and IV. Study the materials carefully in these important phases—they will help you polish your examination-taking skills. Approximately two to three days before you take your next examination, carefully read all the highlighted information in the reference materials using the same techniques you applied during the feedback part. This will reinforce your learning and provide you with an added level of confidence going into the examination.

Someone once said to professional golfer Tom Watson after he won several tournament championships, "You are really lucky to have won those championships. You are really on a streak." Watson was reported to have replied, "Yes, there is some luck involved, but what I've really noticed is that the more I practice, the luckier I get." What Watson was saying is that good luck usually results from good preparation. This line of thinking certainly applies to learning the rules and hints of examination taking.

——————— Rule 7 ———————

Good luck = good preparation.

1. Reference: NFPA 1001, 6.1.1 and 6.1.2
   Delmar, *Firefighter's Handbook*, 3rd Edition, 1st Printing, page 37.
   IFSTA, *Essentials of Fire Fighting and Fire Department Operations*, 5th Edition, 1st Printing, page 33.
   Jones and Bartlett, NFPA, *Fundamentals of Fire Fighter Skills*, 2nd Edition, 1st Printing, page 6.
   Answer: B

2. Reference: NFPA 1001, 6.1.1

Delmar, *Firefighter's Handbook*, 3rd Edition, 1st Printing, page 37.

IFSTA, *Essentials of Fire Fighting and Fire Department Operations*, 5th Edition, 1st Printing, page 33.

Jones and Bartlett, NFPA, *Fundamentals of Fire Fighter Skills*, 2nd Edition, 1st Printing, page 6.

Answer: B

3. Reference: NFPA 1001, 6.1.1 and 6.1.2

Delmar, *Firefighter's Handbook*, 3rd Edition, 1st Printing, page 42.

IFSTA, *Essentials of Fire Fighting and Fire Department Operations*, 5th Edition, 1st Printing, page 37.

Jones and Bartlett, NFPA, *Fundamentals of Fire Fighter Skills*, 2nd Edition, 1st Printing, page 110.

Answer: B

4. Reference: NFPA 1001, 6.1.1 and 6.1.2

Delmar, *Firefighter's Handbook*, 3rd Edition, 1st Printing, page 41.

IFSTA, *Essentials of Fire Fighting and Fire Department Operations*, 5th Edition, 1st Printing, pages 35–36.

Jones and Bartlett, NFPA, *Fundamentals of Fire Fighter Skills*, 2nd Edition, 1st Printing, pages 107–108.

Answer: D

5. Reference: NFPA 1001, 6.1.1 and 6.1.2

Delmar, *Firefighter's Handbook*, 3rd Edition, 1st Printing, page 39.

IFSTA, *Essentials of Fire Fighting and Fire Department Operations*, 5th Edition, 1st Printing, page 70.

Jones and Bartlett, NFPA, *Fundamentals of Fire Fighter Skills*, 2nd Edition, 1st Printing, page 118.

Answer: B

6. Reference: NFPA 1001, 6.1.1 and 6.1.2

Delmar, *Firefighter's Handbook*, 3rd Edition, 1st Printing, pages 189, 824–825.

IFSTA, *Essentials of Fire Fighting and Fire Department Operations*, 5th Edition, 1st Printing, page 74.

Jones and Bartlett, NFPA, *Fundamentals of Fire Fighter Skills*, 2nd Edition, 1st Printing, page 32.

Answer: A

7. Reference: NFPA 1001, 6.1.1 and 6.1.2

Delmar, *Firefighter's Handbook*, 3rd Edition, 1st Printing, page 46.

IFSTA, *Essentials of Fire Fighting and Fire Department Operations*, 5th Edition, 1st Printing, page 39.

Jones and Bartlett, NFPA, *Fundamentals of Fire Fighter Skills*, 2nd Edition, 1st Printing, page 112.

Answer: B

8. Reference: NFPA 1001, 6.1.1 and 6.1.2

Delmar, *Firefighter's Handbook*, 3rd Edition, 1st Printing, pages 44–46.

IFSTA, *Essentials of Fire Fighting and Fire Department Operations*, 5th Edition, 1st Printing, page 36.

Jones and Bartlett, NFPA, *Fundamentals of Fire Fighter Skills*, 2nd Edition, 1st Printing, page 109.

Answer: D

9. Reference: NFPA 1001, 6.1.1 and 6.1.2

Delmar, *Firefighter's Handbook*, 3rd Edition, 1st Printing, page 40.

IFSTA, *Essentials of Fire Fighting and Fire Department Operations*, 5th Edition, 1st Printing, page 35.

Jones and Bartlett, NFPA, *Fundamentals of Fire Fighter Skills*, 2nd Edition, 1st Printing, page 105.

Answer: B

10. Reference: NFPA 1001, 6.1.1 and 6.1.2

Delmar, *Firefighter's Handbook*, 3rd Edition, 1st Printing, page 46.

IFSTA, *Essentials of Fire Fighting and Fire Department Operations*, 5th Edition, 1st Printing, page 36.

Jones and Bartlett, NFPA, *Fundamentals of Fire Fighter Skills*, 2nd Edition, 1st Printing, page 110.

Answer: D

11. Reference: NFPA 1001, 6.1.1 and 6.1.2

Delmar, Fire fighters' Handbook, 3rd Edition, 1st Printing, pages 43 and 44.

IFSTA, *Essentials of Fire Fighting and Fire Department Operations*, 5th Edition, 1st Printing, pages 37–38.

Jones and Bartlett, NFPA, *Fundamentals of Fire Fighter Skills*, 2nd Edition, 1st Printing, page 115.

Answer: B

12. Reference: NFPA 1001, 6.1.1 and 6.1.2

Delmar, *Firefighter's Handbook*, 3rd Edition, 1st Printing, page 43.

IFSTA, *Essentials of Fire Fighting and Fire Department Operations*, 5th Edition, 1st Printing, page 37.

Jones and Bartlett, NFPA, *Fundamentals of Fire Fighter Skills*, 2nd Edition, 1st Printing, page 111.

Answer: B

13. Reference: NFPA 1001, 6.1.1 and 6.1.2

Delmar, *Firefighter's Handbook*, 3rd Edition, 1st Printing, page 43.

IFSTA, *Essentials of Fire Fighting and Fire Department Operations*, 5th Edition, 1st Printing, page 37.

Jones and Bartlett, NFPA, *Fundamentals of Fire Fighter Skills*, 2nd Edition, 1st Printing, page 111.

Answer: A

14. Reference: NFPA 1001, 6.1.1 and 6.1.2
Delmar, *Firefighter's Handbook*, 3rd Edition, 1st Printing, page 46.
IFSTA, *Essentials of Fire Fighting and Fire Department Operations*, 5th Edition, 1st Printing, page 39.
Jones and Bartlett, NFPA, *Fundamentals of Fire Fighter Skills*, 2nd Edition, 1st Printing, page 112.
Answer: A

15. Reference: NFPA 1001, 6.1.1 and 6.1.2
Delmar, *Firefighter's Handbook*, 3rd Edition, 1st Printing, page 41.
IFSTA, *Essentials of Fire Fighting and Fire Department Operations*, 5th Edition, 1st Printing, page 19.
Jones and Bartlett, NFPA, *Fundamentals of Fire Fighter Skills*, 2nd Edition, 1st Printing, page 107.
Answer: B

16. Reference: NFPA 1001, 6.1.1 and 6.1.2
Delmar, *Firefighter's Handbook*, 3rd Edition, 1st Printing, page 76.
IFSTA, *Essentials of Fire Fighting and Fire Department Operations*, 5th Edition, 1st Printing, pages 945–946.
Jones and Bartlett, NFPA, *Fundamentals of Fire Fighter Skills*, 2nd Edition, 1st Printing, page 118.
Answer: D

17. Reference: NFPA 1001, 6.2.2, 6.2.2(A)(B), 6.3.2 and 6.3.2(A)(B)
Delmar, *Firefighter's Handbook*, 3rd Edition, 1st Printing, page 835.
IFSTA, *Essentials of Fire Fighting and Fire Department Operations*, 5th Edition, 1st Printing, page 948.
Jones and Bartlett, NFPA, *Fundamentals of Fire Fighter Skills*, 2nd Edition, 1st Printing, page 96.
Answer: D

18. Reference: NFPA 1001, 6.2.2 and 6.2.2(A)(B)
Delmar, *Firefighter's Handbook*, 3rd Edition, 1st Printing, pages 65–66.
IFSTA, *Essentials of Fire Fighting and Fire Department Operations*, 5th Edition, 1st Printing, page 932.
Jones and Bartlett, NFPA, *Fundamentals of Fire Fighter Skills*, 2nd Edition, 1st Printing, page 84.
Answer: B

19. Reference: NFPA 1001, 6.2.2 and 6.2.2(A)(B)
Delmar, *Firefighter's Handbook*, 3rd Edition, 1st Printing, page 61.
IFSTA, *Essentials of Fire Fighting and Fire Department Operations*, 5th Edition, 1st Printing, page 938.
Jones and Bartlett, NFPA, *Fundamentals of Fire Fighter Skills*, 2nd Edition, 1st Printing, pages 83–85.
Answer: A

20. Reference: NFPA 1001, 6.2.2 and 6.2.2(A)(B)

Delmar, *Firefighter's Handbook*, 3rd Edition, 1st Printing, page 58.

IFSTA, *Essentials of Fire Fighting and Fire Department Operations*, 5th Edition, 1st Printing, page 934.

Jones and Bartlett, NFPA, *Fundamentals of Fire Fighter Skills*, 2nd Edition, 1st Printing, page 81.

Answer: A

21. Reference: NFPA 1001, 6.2.2 and 6.2.2(A)(B)

Delmar, *Firefighter's Handbook*, 3rd Edition, 1st Printing, page 76.

IFSTA, *Essentials of Fire Fighting and Fire Department Operations*, 5th Edition, 1st Printing, page 946.

Jones and Bartlett, NFPA, *Fundamentals of Fire Fighter Skills*, 2nd Edition, 1st Printing, page 283.

Answer: B

22. Reference: NFPA 1001, 6.2.2 and 6.2.2(A)(B)

Delmar, *Firefighter's Handbook*, 3rd Edition, 1st Printing, page 836.

IFSTA, *Essentials of Fire Fighting and Fire Department Operations*, 5th Edition, 1st Printing, page 949.

Jones and Bartlett, NFPA, *Fundamentals of Fire Fighter Skills*, 2nd Edition, 1st Printing, page 535.

Answer: D

23. Reference: NFPA 1001, 6.3.1 and 6.3.1(A)(B)

Delmar, *Firefighter's Handbook*, 3rd Edition, 1st Printing, page 340.

IFSTA, *Essentials of Fire Fighting and Fire Department Operations*, 5th Edition, 1st Printing, page 736.

Jones and Bartlett, NFPA, *Fundamentals of Fire Fighter Skills*, 2nd Edition, 1st Printing, page 521.

Answer: C

24. Reference: NFPA 1001, 6.3.1 and 6.3.1(A)(B)

Delmar, *Firefighter's Handbook*, 3rd Edition, 1st Printing, page 339.

IFSTA, *Essentials of Fire Fighting and Fire Department Operations*, 5th Edition, 1st Printing, page 738.

Jones and Bartlett, NFPA, *Fundamentals of Fire Fighter Skills*, 2nd Edition, 1st Printing, page 522.

Answer: C

25. Reference: NFPA 1001, 6.3.1 and 6.3.1(A)(B)

Delmar, *Firefighter's Handbook*, 3rd Edition, 1st Printing, pages 345–346.

IFSTA, *Essentials of Fire Fighting and Fire Department Operations*, 5th Edition, 1st Printing, page 747.

Jones and Bartlett, NFPA, *Fundamentals of Fire Fighter Skills*, 2nd Edition, 1st Printing, page 523.

Answer: C

26. Reference: NFPA 1001, 6.3.1 and 6.3.1(A)(B)
Delmar, *Firefighter's Handbook*, 3rd Edition, 1st Printing, page 105.
IFSTA, *Essentials of Fire Fighting and Fire Department Operations*, 5th Edition,
1st Printing, page 774.
Jones and Bartlett, NFPA, *Fundamentals of Fire Fighter Skills*, 2nd Edition,
1st Printing, page 633.
Answer: A

27. References: NFPA 1001, 6.3.1 and 6.3.1(A)
Delmar, *Firefighter's Handbook*, 3rd Edition, 1st Printing, page 338.
IFSTA, *Essentials of Fire Fighting and Fire Department Operations*, 5th Edition,
1st Printing, page 734.
Jones and Bartlett, NFPA, *Fundamentals of Fire Fighter Skills*, 2nd Edition,
1st Printing, page 521.
Answer: D

28. Reference: NFPA 1001, 6.3.2 and 6.3.2(A)(B)
Delmar, *Firefighter's Handbook*, 3rd Edition, 1st Printing, page 708.
IFSTA, *Essentials of Fire Fighting and Fire Department Operations*, 5th Edition,
1st Printing, page 783.
Jones and Bartlett, NFPA, *Fundamentals of Fire Fighter Skills*, 2nd Edition,
1st Printing, pages 636–637.
Answer: D

29. Reference: NFPA 1001, 6.3.2 and 6.3.2(A)(B)
Delmar, *Firefighter's Handbook*, 3rd Edition, 1st Printing, pages 393–394.
IFSTA, *Essentials of Fire Fighting and Fire Department Operations*, 5th Edition,
1st Printing, page 146.
Jones and Bartlett, NFPA, *Fundamentals of Fire Fighter Skills*, 2nd Edition,
1st Printing, page 157.
Answer: A

30. Reference: NFPA 1001, 6.3.2 and 6.3.2(A)(B)
Delmar, *Firefighter's Handbook*, 3rd Edition, 1st Printing, page 398.
IFSTA, *Essentials of Fire Fighting and Fire Department Operations*, 5th Edition,
1st Printing, page 149.
Jones and Bartlett, NFPA, *Fundamentals of Fire Fighter Skills*, 2nd Edition,
1st Printing, page 160.
Answer: D

31. Reference: NFPA 1001, 6.3.2 and 6.3.2(A)(B)
Delmar, *Firefighter's Handbook*, 3rd Edition, 1st Printing, page 397.
IFSTA, *Essentials of Fire Fighting and Fire Department Operations*, 5th Edition,
1st Printing, page 148.
Jones and Bartlett, NFPA, *Fundamentals of Fire Fighter Skills*, 2nd Edition,
1st Printing, page 159.
Answer: C

32. Reference: NFPA 1001, 6.3.2 and 6.3.2(A)(B)

Delmar, *Firefighter's Handbook*, 3rd Edition, 1st Printing, page 597.

IFSTA, *Essentials of Fire Fighting and Fire Department Operations*, 5th Edition, 1st Printing, page 409.

Jones and Bartlett, NFPA, *Fundamentals of Fire Fighter Skills*, 2nd Edition, 1st Printing, page 294.

Answer: D

33. Reference: NFPA 1001, 6.3.2 and 6.3.2(A)(B)

Delmar, *Firefighter's Handbook*, 3rd Edition, 1st Printing, pages 744–745.

IFSTA, *Essentials of Fire Fighting and Fire Department Operations*, 5th Edition, 1st Printing, page 884.

Jones and Bartlett, NFPA, *Fundamentals of Fire Fighter Skills*, 2nd Edition, 1st Printing, page 576.

Answer: D

34. Reference: NFPA 1001, 6.3.2 and 6.3.2(A)(B)

Delmar, *Firefighter's Handbook*, 3rd Edition, 1st Printing, pages 90–91.

IFSTA, *Essentials of Fire Fighting and Fire Department Operations*, 5th Edition, 1st Printing, page 93.

Jones and Bartlett, NFPA, *Fundamentals of Fire Fighter Skills*, 2nd Edition, 1st Printing, page 129.

Answer: C

35. Reference: NFPA 1001, 6.3.2 and 6.3.2(A)(B)

Delmar, *Firefighter's Handbook*, 3rd Edition, 1st Printing, page 656.

IFSTA, *Essentials of Fire Fighting and Fire Department Operations*, 5th Edition, 1st Printing, page 544.

Jones and Bartlett, NFPA, *Fundamentals of Fire Fighter Skills*, 2nd Edition, 1st Printing, page 424.

Answer: B

36. Reference: NFPA 1001, 6.3.2 and 6.3.2(A)(B)

Delmar, *Firefighter's Handbook*, 3rd Edition, 1st Printing, page 655.

IFSTA, *Essentials of Fire Fighting and Fire Department Operations*, 5th Edition, 1st Printing, page 577.

Jones and Bartlett, NFPA, *Fundamentals of Fire Fighter Skills*, 2nd Edition, 1st Printing, pages 416–417.

Answer: A

37. Reference: NFPA 1001, 6.3.2 and 6.3.2(A)(B)

Delmar, *Firefighter's Handbook*, 3rd Edition, 1st Printing, page 362.

IFSTA, *Essentials of Fire Fighting and Fire Department Operations*, 5th Edition, 1st Printing, page 856.

Jones and Bartlett, NFPA, *Fundamentals of Fire Fighter Skills*, 2nd Edition, 1st Printing, page 960.

Answer: B

38. Reference: NFPA 1001, 6.3.2 and 6.3.2(A)(B)
Delmar, *Firefighter's Handbook*, 3rd Edition, 1st Printing, page 705.
IFSTA, *Essentials of Fire Fighting and Fire Department Operations*, 5th Edition,
1st Printing, page 793.
Jones and Bartlett, NFPA, *Fundamentals of Fire Fighter Skills*, 2nd Edition,
1st Printing, page 628.
Answer: B

39. Reference: NFPA 1001, 6.3.2 and 6.3.2(A)(B)
Delmar, *Firefighter's Handbook*, 3rd Edition, 1st Printing, page 121.
IFSTA, *Essentials of Fire Fighting and Fire Department Operations*, 5th Edition,
1st Printing, page 760.
Jones and Bartlett, NFPA, *Fundamentals of Fire Fighter Skills*, 2nd Edition,
1st Printing, page 31.
Answer: A

40. Reference: NFPA 1001, 6.3.2 and 6.3.2(A)(B)
Delmar, *Firefighter's Handbook*, 3rd Edition, 1st Printing, page 660.
IFSTA, *Essentials of Fire Fighting and Fire Department Operations*, 5th Edition,
1st Printing, page 567.
Jones and Bartlett, NFPA, *Fundamentals of Fire Fighter Skills*, 2nd Edition,
1st Printing, page 432.
Answer: A

41. Reference: NFPA 1001, 6.3.2 and 6.3.2(A)(B)
Delmar, *Firefighter's Handbook*, 3rd Edition, 1st Printing, pages 141 and 151.
IFSTA, *Essentials of Fire Fighting and Fire Department Operations*, 5th Edition,
1st Printing, page 881.
Jones and Bartlett, NFPA, *Fundamentals of Fire Fighter Skills*, 2nd Edition,
1st Printing, page 573.
Answer: C

42. References: NFPA 1001, 6.3.2 and 6.3.2(A)(B)
Delmar, *Firefighter's Handbook*, 3rd Edition, 1st Printing, page 391.
IFSTA, *Essentials of Fire Fighting and Fire Department Operations*, 5th Edition,
1st Printing, page 142.
Jones and Bartlett, NFPA, *Fundamentals of Fire Fighter Skills*, 2nd Edition,
1st Printing, pages 153–154.
Answer: D

43. Reference: NFPA 1001, 6.3.3 and 6.3.3(A)(B)
Delmar, *Firefighter's Handbook*, 3rd Edition, 1st Printing, page 890.
IFSTA, *Essentials of Fire Fighting and Fire Department Operations*, 5th Edition,
1st Printing, page 774.
Jones and Bartlett, NFPA, *Fundamentals of Fire Fighter Skills*, 2nd Edition,
1st Printing, page 635.
Answer: B

44. References: NFPA 1001, 6.3.3 and 6.3.3(A)

Delmar, *Firefighter's Handbook*, 3rd Edition, 1st Printing, page 908.

IFSTA, *Essentials of Fire Fighting and Fire Department Operations*, 5th Edition, 1st Printing, page 779.

Jones and Bartlett, NFPA, *Fundamentals of Fire Fighter Skills*, 2nd Edition, 1st Printing, page 141.

Answer: B

45. Reference: NFPA 1001, 6.3.3 and 6.3.3(A)(B)

Delmar, *Firefighter's Handbook*, 3rd Edition, 1st Printing, page 908.

IFSTA, *Essentials of Fire Fighting and Fire Department Operations*, 5th Edition, 1st Printing, page 779.

Jones and Bartlett, NFPA, *Fundamentals of Fire Fighter Skills*, 2nd Edition, 1st Printing, page 141.

Answer: C

46. Reference: NFPA 1001, 6.3.4 and 6.3.4(A)(B)

Delmar, *Firefighter's Handbook*, 3rd Edition, 1st Printing, page 746.

IFSTA, *Essentials of Fire Fighting and Fire Department Operations*, 5th Edition, 1st Printing, pages 911–913.

Jones and Bartlett, NFPA, *Fundamentals of Fire Fighter Skills*, 2nd Edition, 1st Printing, pages 981–983.

Answer: D

47. Reference: NFPA 1001, 6.3.4 and 6.3.4(A)(B)

Delmar, *Firefighter's Handbook*, 3rd Edition, 1st Printing, page 747.

IFSTA, *Essentials of Fire Fighting and Fire Department Operations*, 5th Edition, 1st Printing, page 918.

Jones and Bartlett, NFPA, *Fundamentals of Fire Fighter Skills*, 2nd Edition, 1st Printing, page 977.

Answer: D

48. Reference: NFPA 1001, 6.3.4 and 6.3.4(A)(B)

Delmar, *Firefighter's Handbook*, 3rd Edition, 1st Printing, page 747.

IFSTA, *Essentials of Fire Fighting and Fire Department Operations*, 5th Edition, 1st Printing, page 918.

Jones and Bartlett, NFPA, *Fundamentals of Fire Fighter Skills*, 2nd Edition, 1st Printing, page 977.

Answer: D

49. Reference: NFPA 1001, 6.4.1 and 6.4.1(A)(B)

Delmar, *Firefighter's Handbook*, 3rd Edition, 1st Printing, page 553.

IFSTA, *Essentials of Fire Fighting and Fire Department Operations*, 5th Edition, 1st Printing, page 352.

Jones and Bartlett, NFPA, *Fundamentals of Fire Fighter Skills*, 2nd Edition, 1st Printing, page 729.

Answer: C

50. Reference: NFPA 1001, 6.4.1 and 6.4.1(A)(B)

Delmar, *Firefighter's Handbook*, 3rd Edition, 1st Printing, page 557.

IFSTA, *Essentials of Fire Fighting and Fire Department Operations*, 5th Edition, 1st Printing, page 356.

Jones and Bartlett, NFPA, *Fundamentals of Fire Fighter Skills*, 2nd Edition, 1st Printing, page 754.

Answer: B

51. Reference: NFPA 1001, 6.4.1 and 6.4.1(A)(B)

Delmar, *Firefighter's Handbook*, 3rd Edition, 1st Printing, pages 555–556.

IFSTA, *Essentials of Fire Fighting and Fire Department Operations*, 5th Edition, 1st Printing, page 358.

Jones and Bartlett, NFPA, *Fundamentals of Fire Fighter Skills*, 2nd Edition, 1st Printing, page 735.

Answer: B

52. Reference: NFPA 1001, 6.4.1 and 6.4.1(A)

Delmar, *Firefighter's Handbook*, 3rd Edition, 1st Printing, page 555.

IFSTA, *Essentials of Fire Fighting and Fire Department Operations*, 5th Edition, 1st Printing, page 357.

Jones and Bartlett, NFPA, *Fundamentals of Fire Fighter Skills*, 2nd Edition, 1st Printing, page 735.

Answer: A

53. Reference: NFPA 1001, 6.4.1 and 6.4.1(A)(B)

Delmar, *Firefighter's Handbook*, 3rd Edition, 1st Printing, page 557.

IFSTA, *Essentials of Fire Fighting and Fire Department Operations*, 5th Edition, 1st Printing, pages 355–356.

Jones and Bartlett, NFPA, *Fundamentals of Fire Fighter Skills*, 2nd Edition, 1st Printing, page 743.

Answer: B

54. Reference: NFPA 1001, 6.4.1, 6.4.1(A)(B), 6.4.2, and 6.4.2(A)(B)

Delmar, *Firefighter's Handbook*, 3rd Edition, 1st Printing, page 550.

IFSTA, *Essentials of Fire Fighting and Fire Department Operations*, 5th Edition, 1st Printing, page 345.

Jones and Bartlett, NFPA, *Fundamentals of Fire Fighter Skills*, 2nd Edition, 1st Printing, page 731.

Answer: D

55. Reference: NFPA 1001, 6.4.1 and 6.4.1(A)

Delmar, *Firefighter's Handbook*, 3rd Edition, 1st Printing, pages 551–552.

IFSTA, *Essentials of Fire Fighting and Fire Department Operations*, 5th Edition, 1st Printing, page 72.

Jones and Bartlett, NFPA, *Fundamentals of Fire Fighter Skills*, 2nd Edition, 1st Printing, page 726.

Answer: C

56. Reference: NFPA 1001, 6.4.1 and 6.4.1(A)(B)
Delmar, *Firefighter's Handbook*, 3rd Edition, 1st Printing, page 550.
IFSTA, *Essentials of Fire Fighting and Fire Department Operations*, 5th Edition, 1st Printing, page 345.
Jones and Bartlett, NFPA, *Fundamentals of Fire Fighter Skills*, 2nd Edition, 1st Printing, page 731.
Answer: A

57. Reference: NFPA 1001, 6.4.1 and 6.4.1(A)(B)
Delmar, *Firefighter's Handbook*, 3rd Edition, 1st Printing, page 555.
IFSTA, *Essentials of Fire Fighting and Fire Department Operations*, 5th Edition, 1st Printing, page 357.
Jones and Bartlett, NFPA, *Fundamentals of Fire Fighter Skills*, 2nd Edition, 1st Printing, page 740.
Answer: A

58. Reference: NFPA 1001, 6.4.2 and 6.4.2(A)(B)
Delmar, *Firefighter's Handbook*, 3rd Edition, 1st Printing, page 550.
IFSTA, *Essentials of Fire Fighting and Fire Department Operations*, 5th Edition, 1st Printing, page 345.
Jones and Bartlett, NFPA, *Fundamentals of Fire Fighter Skills*, 2nd Edition, 1st Printing, page 731.
Answer: A

59. Reference: NFPA 1001, 6.4.2 and 6.4.2(A)(B)
Delmar, *Firefighter's Handbook*, 3rd Edition, 1st Printing, page 570.
IFSTA, *Essentials of Fire Fighting and Fire Department Operations*, 5th Edition, 1st Printing, page 365.
Jones and Bartlett, NFPA, *Fundamentals of Fire Fighter Skills*, 2nd Edition, 1st Printing, page 763.
Answer: A

60. Reference: NFPA 1001, 6.4.2 and 6.4.2(A)(B)
Delmar, *Firefighter's Handbook*, 3rd Edition, 1st Printing, page 573.
IFSTA, *Essentials of Fire Fighting and Fire Department Operations*, 5th Edition, 1st Printing, pages 367 and 800.
Jones and Bartlett, NFPA, *Fundamentals of Fire Fighter Skills*, 2nd Edition, 1st Printing, page 759.
Answer: A

61. Reference: NFPA 1001, 6.4.2 and 6.4.2(A)(B)
Delmar, *Firefighter's Handbook*, 3rd Edition, 1st Printing, page 555.
IFSTA, *Essentials of Fire Fighting and Fire Department Operations*, 5th Edition, 1st Printing, page 355.
Jones and Bartlett, NFPA, *Fundamentals of Fire Fighter Skills*, 2nd Edition, 1st Printing, page 754.
Answer: D

62. Reference: NFPA 1001, 6.5.1 and 6.5.1(A)(B)

Delmar, *Firefighter's Handbook*, 3rd Edition, 1st Printing, page 391.

IFSTA, *Essentials of Fire Fighting and Fire Department Operations*, 5th Edition, 1st Printing, page 144.

Jones and Bartlett, NFPA, *Fundamentals of Fire Fighter Skills*, 2nd Edition, 1st Printing, page 154.

Answer: B

63. Reference: NFPA 1001, 6.5.1, 6.5.1(A), 6.5.2, 6.5.2(A)(B), 6.5.3, and 6.5.3(A)(B)

Delmar, *Firefighter's Handbook*, 3rd Edition, 1st Printing, page 754.

IFSTA, *Essentials of Fire Fighting and Fire Department Operations*, 5th Edition, 1st Printing, pages 974 and 959.

Jones and Bartlett, NFPA, *Fundamentals of Fire Fighter Skills*, 2nd Edition, 1st Printing, page 918.

Answer: A

64. References: NFPA 1001, 6.5.1 and 6.5.1(A)(B)

Delmar, *Firefighter's Handbook*, 3rd Edition, 1st Printing, page 395.

IFSTA, *Essentials of Fire Fighting and Fire Department Operations*, 5th Edition, 1st Printing, pages 147.

Jones and Bartlett, NFPA, *Fundamentals of Fire Fighter Skills*, 2nd Edition, 1st Printing, page 648.

Answer: B

65. Reference: NFPA 1001, 6.5.1 and 6.5.1(A)

Delmar, *Firefighter's Handbook*, 3rd Edition, 1st Printing, page 782.

IFSTA, *Essentials of Fire Fighting and Fire Department Operations*, 5th Edition, 1st Printing, page 966.

Jones and Bartlett, NFPA, *Fundamentals of Fire Fighter Skills*, 2nd Edition, 1st Printing, page 644.

Answer: C

66. Reference: NFPA 1001, 6.5.1 and 6.5.1(A)(B)

Delmar, *Firefighter's Handbook*, 3rd Edition, 1st Printing, pages 775–776.

IFSTA, *Essentials of Fire Fighting and Fire Department Operations*, 5th Edition, 1st Printing, page 977.

Jones and Bartlett, NFPA, *Fundamentals of Fire Fighter Skills*, 2nd Edition, 1st Printing, page 924.

Answer: A

67. Reference: NFPA 1001, 6.5.1 and 6.5.1(A)(B)

Delmar, *Firefighter's Handbook*, 3rd Edition, 1st Printing, page 782.

IFSTA, *Essentials of Fire Fighting and Fire Department Operations*, 5th Edition, 1st Printing, page 961.

Jones and Bartlett, NFPA, *Fundamentals of Fire Fighter Skills*, 2nd Edition, 1st Printing, page 658.

Answer: D

68. Reference: NFPA 1001, 6.5.1 and 6.5.1(A)(B)

Delmar, *Firefighter's Handbook*, 3rd Edition, 1st Printing, page 775.

IFSTA, *Essentials of Fire Fighting and Fire Department Operations*, 5th Edition, 1st Printing, page 977.

Jones and Bartlett, NFPA, *Fundamentals of Fire Fighter Skills*, 2nd Edition, 1st Printing, page 922.

Answer: C

69. Reference: NFPA 1001, 6.5.2 and 6.5.2(A)(B)

Delmar, *Firefighter's Handbook*, 3rd Edition, 1st Printing, page 34.

IFSTA, *Essentials of Fire Fighting and Fire Department Operations*, 5th Edition, 1st Printing, page 982.

Jones and Bartlett, NFPA, *Fundamentals of Fire Fighter Skills*, 2nd Edition, 1st Printing, page 918.

Answer: B

70. Reference: NFPA 1001, 6.5.2 and 6.5.2(A)(B)

Delmar, *Firefighter's Handbook*, 3rd Edition, 1st Printing, page 780

IFSTA, *Essentials of Fire Fighting and Fire Department Operations*, 5th Edition, 1st Printing, page 987.

Jones and Bartlett, NFPA, *Fundamentals of Fire Fighter Skills*, 2nd Edition, 1st Printing, page 919.

Answer: B

71. Reference: NFPA 1001, 6.5.3 and 6.5.3(A)(B)

Delmar, *Firefighter's Handbook*, 3rd Edition, 1st Printing, page 353.

IFSTA, *Essentials of Fire Fighting and Fire Department Operations*, 5th Edition, 1st Printing, page 827.

Jones and Bartlett, NFPA, *Fundamentals of Fire Fighter Skills*, 2nd Edition, 1st Printing, page 943.

Answer: B

72. Reference: NFPA 1001, 6.5.3 and 6.5.3(A)(B)

Delmar, *Firefighter's Handbook*, 3rd Edition, 1st Printing, page 353.

IFSTA, *Essentials of Fire Fighting and Fire Department Operations*, 5th Edition, 1st Printing, pages 832–833.

Jones and Bartlett, NFPA, *Fundamentals of Fire Fighter Skills*, 2nd Edition, 1st Printing, page 940.

Answer: C

73. Reference: NFPA 1001, 6.5.3 and 6.5.3(A)(B)

Delmar, *Firefighter's Handbook*, 3rd Edition, 1st Printing, page 366.

IFSTA, *Essentials of Fire Fighting and Fire Department Operations*, 5th Edition, 1st Printing, page 850.

Jones and Bartlett, NFPA, *Fundamentals of Fire Fighter Skills*, 2nd Edition, 1st Printing, page 957.

Answer: A

74. Reference: NFPA 1001, 6.5.3 and 6.5.3(A)(B)
Delmar, *Firefighter's Handbook*, 3rd Edition, 1st Printing, pages 367–368.
IFSTA, *Essentials of Fire Fighting and Fire Department Operations*, 5th Edition,
1st Printing, page 848.
Jones and Bartlett, NFPA, *Fundamentals of Fire Fighter Skills*, 2nd Edition,
1st Printing, page 954.
Answer: B

75. Reference: NFPA 1001, 6.5.3 and 6.5.3(A)(B)
Delmar, *Firefighter's Handbook*, 3rd Edition, 1st Printing, page 361.
IFSTA, *Essentials of Fire Fighting and Fire Department Operations*, 5th Edition,
1st Printing, page 851.
Jones and Bartlett, NFPA, *Fundamentals of Fire Fighter Skills*, 2nd Edition,
1st Printing, page 956.
Answer: B

76. Reference: NFPA 1001, 6.5.3 and 6.5.3(A)(B)
Delmar, *Firefighter's Handbook*, 2nd Edition, 1st Printing, page 237.
IFSTA, *Essentials of Fire Fighting and Fire Department Operations*, 5th Edition,
1st Printing, page 595.
Jones and Bartlett, NFPA, *Fundamentals of Fire Fighter Skills*, 2nd Edition,
1st Printing, page 449.
Answer: C

77. Reference: NFPA 1001, 6.5.3 and 6.5.3(A)(B)
Delmar, *Firefighter's Handbook*, 3rd Edition, 1st Printing, page 238.
IFSTA, *Essentials of Fire Fighting and Fire Department Operations*, 5th Edition,
1st Printing, page 599.
Jones and Bartlett, NFPA, *Fundamentals of Fire Fighter Skills*, 2nd Edition,
1st Printing, page 450.
Answer: D

78. Reference: NFPA 1001, 6.5.3 and 6.5.3(A)(B)
Delmar, *Firefighter's Handbook*, 3rd Edition, 1st Printing, page 239.
IFSTA, *Essentials of Fire Fighting and Fire Department Operations*, 5th Edition,
1st Printing, page 606.
Jones and Bartlett, NFPA, *Fundamentals of Fire Fighter Skills*, 2nd Edition,
1st Printing, page 453.
Answer: C

79. Reference: NFPA 1001, 6.5.3 and 6.5.3(A)(B)
Delmar, *Firefighter's Handbook*, 3rd Edition, 1st Printing, page 238.
IFSTA, *Essentials of Fire Fighting and Fire Department Operations*, 5th Edition,
1st Printing, page 608.
Jones and Bartlett, NFPA, *Fundamentals of Fire Fighter Skills*, 2nd Edition,
1st Printing, page 452.
Answer: B

80. Reference: NFPA 1001, 6.5.3 and 6.5.3(A)(B)

Delmar, *Firefighter's Handbook*, 3rd Edition, 1st Printing, page 238.

IFSTA, *Essentials of Fire Fighting and Fire Department Operations*, 5th Edition, 1st Printing, pages 599–600.

Jones and Bartlett, NFPA, *Fundamentals of Fire Fighter Skills*, 2nd Edition, 1st Printing, page 450.

Answer: D

81. Reference: NFPA 1001, 6.5.3 and 6.5.3(A)(B)

Delmar, *Firefighter's Handbook*, 3rd Edition, 1st Printing, page 238.

IFSTA, *Essentials of Fire Fighting and Fire Department Operations*, 5th Edition, 1st Printing, page 599.

Jones and Bartlett, NFPA, *Fundamentals of Fire Fighter Skills*, 2nd Edition, 1st Printing, page 450.

Answer: C

82. Reference: NFPA 1001, 6.5.3 and 6.5.3(A)(B)

Delmar, *Firefighter's Handbook*, 3rd Edition, 1st Printing, pages 237–238.

IFSTA, *Essentials of Fire Fighting and Fire Department Operations*, 5th Edition, 1st Printing, page 599.

Jones and Bartlett, NFPA, *Fundamentals of Fire Fighter Skills*, 2nd Edition, 1st Printing, page 450.

Answer: A

83. Reference: NFPA 1001, 6.5.3 and 6.5.3(A)(B)

Delmar, *Firefighter's Handbook*, 3rd Edition, 1st Printing, page 238.

IFSTA, *Essentials of Fire Fighting and Fire Department Operations*, 5th Edition, 1st Printing, page 606.

Jones and Bartlett, NFPA, *Fundamentals of Fire Fighter Skills*, 2nd Edition, 1st Printing, page 452.

Answer: A

84. Reference: NFPA 1001, 6.5.3 and 6.5.3(A)(B)

Delmar, *Firefighter's Handbook*, 3rd Edition, 1st Printing, pages 237–238.

IFSTA, *Essentials of Fire Fighting and Fire Department Operations*, 5th Edition, 1st Printing, page 599.

Jones and Bartlett, NFPA, *Fundamentals of Fire Fighter Skills*, 2nd Edition, 1st Printing, page 450.

Answer: B

85. Reference: NFPA 1001, 6.5.3 and 6.5.3(A)(B)

Delmar, *Firefighter's Handbook*, 3rd Edition, 1st Printing, page 365.

IFSTA, *Essentials of Fire Fighting and Fire Department Operations*, 5th Edition, 1st Printing, page 855.

Jones and Bartlett, NFPA, *Fundamentals of Fire Fighter Skills*, 2nd Edition, 1st Printing, pages 960–961.

Answer: A

86. Reference: NFPA 1001, 6.5.3 and 6.5.3(A)(B)

Delmar, *Firefighter's Handbook*, 3rd Edition, 1st Printing, page 367.

IFSTA, *Essentials of Fire Fighting and Fire Department Operations*, 5th Edition, 1st Printing, page 848.

Jones and Bartlett, NFPA, *Fundamentals of Fire Fighter Skills*, 2nd Edition, 1st Printing, page 954.

Answer: C

87. Reference: NFPA 1001, 6.5.3 and 6.5.3(A)(B)

Delmar, *Firefighter's Handbook*, 3rd Edition, 1st Printing, page 363.

IFSTA, *Essentials of Fire Fighting and Fire Department Operations*, 5th Edition, 1st Printing, page 853.

Jones and Bartlett, NFPA, *Fundamentals of Fire Fighter Skills*, 2nd Edition, 1st Printing, page 958.

Answer: D

88. Reference: NFPA 1001, 6.5.3 and 6.5.3(A)(B)

Delmar, *Firefighter's Handbook*, 2nd Edition, 1st Printing, page 247.

IFSTA, *Essentials of Fire Fighting and Fire Department Operations*, 5th Edition, 1st Printing, pages 604, 608–609.

Jones and Bartlett, NFPA, *Fundamentals of Fire Fighter Skills*, 2nd Edition, 1st Printing, pages 455–457.

Answer: D

89. Reference: NFPA 1001, 6.5.3 and 6.5.3(A)(B)

Delmar, *Firefighter's Handbook*, 3rd Edition, 1st Printing, page 361.

IFSTA, *Essentials of Fire Fighting and Fire Department Operations*, 5th Edition, 1st Printing, page 852.

Jones and Bartlett, NFPA, *Fundamentals of Fire Fighter Skills*, 2nd Edition, 1st Printing, page 958.

Answer: A

90. Reference: NFPA 1001, 6.5.3 and 6.5.3(A)(B)

Delmar, *Firefighter's Handbook*, 3rd Edition, 1st Printing, page 363.

IFSTA, *Essentials of Fire Fighting and Fire Department Operations*, 5th Edition, 1st Printing, page 853.

Jones and Bartlett, NFPA, *Fundamentals of Fire Fighter Skills*, 2nd Edition, 1st Printing, page 959.

Answer: C

91. Reference: NFPA 1001, 6.5.3 and 6.5.3(A)(B)

Delmar, *Firefighter's Handbook*, 3rd Edition, 1st Printing, page 361.

IFSTA, *Essentials of Fire Fighting and Fire Department Operations*, 5th Edition, 1st Printing, page 845.

Jones and Bartlett, NFPA, *Fundamentals of Fire Fighter Skills*, 2nd Edition, 1st Printing, page 954.

Answer: D

92. Reference: NFPA 1001, 6.5.3 and 6.5.3(A)(B)

Delmar, *Firefighter's Handbook*, 2nd Edition, 1st Printing, pages 353, 355, and 357.

IFSTA, *Essentials of Fire Fighting and Fire Department Operations*, 5th Edition, 1st Printing, pages 826–834.

Jones and Bartlett, NFPA, *Fundamentals of Fire Fighter Skills*, 2nd Edition, 1st Printing, pages 942–944.

Answer: A

93. Reference: NFPA 1001, 6.5.5 and 6.5.5(A)(B)

Delmar, *Firefighter's Handbook*, 3rd Edition, 1st Printing, page 316.

IFSTA, *Essentials of Fire Fighting and Fire Department Operations*, 5th Edition, 1st Printing, page 713.

Jones and Bartlett, NFPA, *Fundamentals of Fire Fighter Skills*, 2nd Edition, 1st Printing, page 479.

Answer: B

94. Reference: NFPA 1001, 6.5.5 and 6.5.5(A)(B)

Delmar, *Firefighter's Handbook*, 3rd Edition, 1st Printing, page 316.

IFSTA, *Essentials of Fire Fighting and Fire Department Operations*, 5th Edition, 1st Printing, page 713.

Jones and Bartlett, NFPA, *Fundamentals of Fire Fighter Skills*, 2nd Edition, 1st Printing, page 479.

Answer: A

95. Reference: NFPA 1001, 6.5.5 and 6.5.5(A)(B)

Delmar, *Firefighter's Handbook*, 3rd Edition, 1st Printing, page 316.

IFSTA, *Essentials of Fire Fighting and Fire Department Operations*, 5th Edition, 1st Printing, page 713.

Jones and Bartlett, NFPA, *Fundamentals of Fire Fighter Skills*, 2nd Edition, 1st Printing, page 479.

Answer: C

96. Reference: NFPA 1001, 6.5.5 and 6.5.5(A)(B)

Delmar, *Firefighter's Handbook*, 3rd Edition, 1st Printing, page 315.

IFSTA, *Essentials of Fire Fighting and Fire Department Operations*, 5th Edition, 1st Printing, page 680.

Jones and Bartlett, NFPA, *Fundamentals of Fire Fighter Skills*, 2nd Edition, 1st Printing, page 476.

Answer: B

97. Reference: NFPA 1001, 6.5.5 and 6.5.5(A)(B)

Delmar, *Firefighter's Handbook*, 3rd Edition, 1st Printing, page 316.

IFSTA, *Essentials of Fire Fighting and Fire Department Operations*, 5th Edition, 1st Printing, page 682.

Jones and Bartlett, NFPA, *Fundamentals of Fire Fighter Skills*, 2nd Edition, 1st Printing, page 479.

Answer: B

98. Reference: NFPA 1001, 6.5.5 and 6.5.5(A)(B)

Delmar, *Firefighter's Handbook*, 3rd Edition, 1st Printing, page 316.

IFSTA, *Essentials of Fire Fighting and Fire Department Operations*, 5th Edition, 1st Printing, page 713.

Jones and Bartlett, NFPA, *Fundamentals of Fire Fighter Skills*, 2nd Edition, 1st Printing, page 479.

Answer: B

99. Reference: NFPA 1001, 6.5.5 and 6.5.5(A)(B)

Delmar, *Firefighter's Handbook*, 3rd Edition, 1st Printing, page 316.

IFSTA, *Essentials of Fire Fighting and Fire Department Operations*, 5th Edition, 1st Printing, page 713.

Jones and Bartlett, NFPA, *Fundamentals of Fire Fighter Skills*, 2nd Edition, 1st Printing, page 479.

Answer: D

100. Reference: NFPA 1001, 6.5.5 and 6.5.5(A)(B)

Delmar, *Firefighter's Handbook*, 3rd Edition, 1st Printing, page 316.

IFSTA, *Essentials of Fire Fighting and Fire Department Operations*, 5th Edition, 1st Printing, page 680.

Jones and Bartlett, NFPA, *Fundamentals of Fire Fighter Skills*, 2nd Edition, 1st Printing, page 479.

Answer: C

Bibliography for Exam Prep: Fire Fighter I and II, Second Edition

1. National Fire Protection Association, NFPA *1001, Standard for Fire Fighter Professional Qualifications, 2008 Edition.*

2. Delmar, *Firefighter's Handbook, Essentials of Firefighting and Emergency Response,* 3rd Edition, 1st Printing.

3. IFSTA, *Essentials of Fire Fighting and Fire Department Operations,* 5th Edition, 1st Printing.

4. Jones and Bartlett, NFPA, *Fundamentals of Fire Fighter Skills,* 2nd Edition, 1st Printing.

Performance Training Systems, Inc.
Training and testing that are on target!

Online examinations for the Fire and Emergency Medical Services

Registration

FREE OFFER - 150 ITEM PRACTICE TEST - VALUED AT $39.00

Complete registration form and fax it to (561) 277–9402.
PLEASE DO NOT FORGET TO FAX YOUR COMPLETED PERSONAL PROGRESS PLOTTER IN ORDER TO RECEIVE YOUR FREE TEST.

Name

Title

Department

Address: Street

City   State   Zip Code

Telephone   Fax

E-mail

Choose the tests that apply to your needs.

- ❑ Aerial Operator
- ❑ Airport Fire Fighter
- ❑ Confined Space Rescue
- ❑ EMT-Basic
- ❑ Fire and Life Safety Educator I
- ❑ Fire and Life Safety Educator II
- ❑ Fire Department Safety Officer
- ❑ Fire Fighter I
- ❑ Fire Fighter II
- ❑ Fire Inspector I
- ❑ Fire Inspector II

- ❑ Fire Instructor I
- ❑ Fire Instructor II
- ❑ Fire Investigator
- ❑ Fire Officer I
- ❑ Fire Officer II
- ❑ Fire Officer III
- ❑ Fire Officer IV
- ❑ HazMat Awareness
- ❑ HazMat Operations
- ❑ HazMat Technician
- ❑ High Angle Rescue
- ❑ Industrial Fire Fighter-Incipient Level

- ❑ Medical First Responder
- ❑ Paramedic
- ❑ Pumper Driver
- ❑ Ropes and Rigging Rescue
- ❑ Safety Officer
- ❑ Structural Collapse Rescue
- ❑ Swift Water Rescue
- ❑ Vehicle/Machinery Rescue
- ❑ Water/Ice Rescue
- ❑ Wildland Fire Fighter I
- ❑ Wildland Fire Fighter II

Signature:

Order Additional Manuals in the Exam Prep Series Today!

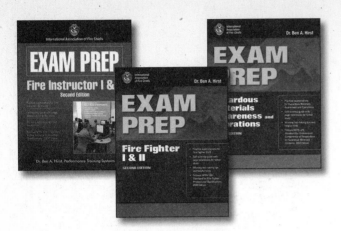

Each manual in the series is designed to prepare you to sit for training, certification, or promotional examinations by including the same type of multiple-choice questions you are likely to encounter on your examination.

The practice examinations were written by fire service personnel and the content was validated through current reference materials and technical review committees. Additionally, each manual includes:

- Self-scoring guide with page references to leading resources for further study
- Winning test-taking tips and helpful hints
- Extensive coverage of relevant NFPA standards
- Free access to a 150-question online practice exam (normally valued at $39.00!)

Manuals within the series are only $29.95* unless otherwise noted. **Volume discounts available.**

Don't waste time or money with other prep resources. Get proven results with the Exam Prep Series.

Visit *www.jbpub.com/Fire/ExamPrep* or call 1-800-832-0034 to place your risk-free order now.

PERFORMANCE TRAINING SYSTEMS, INC. DR. BEN A. HIRST

JONES AND BARTLETT PUBLISHERS
BOSTON   TORONTO   LONDON   SINGAPORE

Exam Prep: Airport Fire Fighter
ISBN-13: 978-0-7637-3764-1

Exam Prep: Building Construction for the Fire Service†
ISBN-13: 978-0-7637-5341-2

Exam Prep: EMT-Basic†
ISBN-13: 978-0-7637-4213-3

Exam Prep: Fire Apparatus Driver Operator†
ISBN-13: 978-0-7637-2845-8

Exam Prep: Fire Department Safety Office†
ISBN-13: 978-0-7637-2846-5

Exam Prep: Fire Fighter I & II, Second Edition†
ISBN-13: 978-0-7637-5836-3

Exam Prep: Fire Inspector I & II†
ISBN-13: 978-0-7637-2848-9

Exam Prep: Fire Instructor I & II, Second Edition†
ISBN-13: 978-0-7637-5837-0

Exam Prep: Fire Investigator†
ISBN-13: 978-0-7637-2849-6

Exam Prep: Fire & Life Safety Educator I & II†
ISBN-13: 978-0-7637-2854-0

Exam Prep: Fire Officer I & II†
ISBN-13: 978-0-7637-2761-1

Exam Prep: Fire Officer III & IV†
ISBN-13: 978-0-7637-4465-6

Exam Prep: Hazardous Materials Awareness & Operations, Second Edition
ISBN-13: 978-0-7637-2853-5838-7

Exam Prep: Hazardous Materials Technician
ISBN-13: 978-0-7637-2852-6

Exam Prep: Industrial Fire-Incipient Level
ISBN-13: 978-0-7637-4212-6

Exam Prep: Medical First Responder†
ISBN-13: 978-0-7637-4214-0

Exam Prep: Paramedic†
ISBN-13: 978-0-7637-4216-4

Exam Prep: Technical Rescue- High Angle
ISBN-13: 978-0-7637-4217-1

Exam Prep: Technical Rescue- Structural Collapse and Confined Space
ISBN-13: 978-0-7637-2906-6

Exam Prep: Technical Rescue- Vehicle/ Machinery and Water/ Ice Rescue
ISBN-13: 978-0-7637-2851-9

Exam Prep: Technical Rescue-Ropes and Rigging
ISBN-13: 978-0-7637-2850-2

Exam Prep: Technical Rescue- Swift Water
ISBN-13: 978-0-7637-5167-8

Exam Prep: Technical Rescue- Trench and Structural Collapse
ISBN-13: 978-0-7637-4218-8

Exam Prep: Telecommunicator I & II†
ISBN-13: 978-0-7637-2856-4

Exam Prep: Wildland Fire Fighter I & II†
ISBN-13: 978-0-7637-2855-7

† $39.95* (Sugg. US List)

www.jbpub.com/Fire